13|50

Sex and
Evolution

MONOGRAPHS IN POPULATION BIOLOGY

EDITED BY ROBERT M. MAY

Sex and Evolution

GEORGE C. WILLIAMS

PRINCETON, NEW JERSEY

PRINCETON UNIVERSITY PRESS

1975

Copyright © 1975 by Princeton University Press
Published by Princeton University Press, Princeton and London

ALL RIGHTS RESERVED

Library of Congress Cataloging in Publication Data will be found
on the last printed page of this book

This book has been composed in Monotype Baskerville
Printed in the United States of America

Preface

This book is written from a conviction that the prevalence of sexual reproduction in higher plants and animals is inconsistent with current evolutionary theory. My purpose is to propose minimal modifications of the theory in order to account for the persistence of so seemingly maladaptive a character. Many well informed readers may disagree with much of my reasoning, but I hope at least to convince them that there is a kind of crisis at hand in evolutionary biology and that my suggestions are plausible enough to warrant serious consideration.

The conceptual complexities and diversity of relevant information on sexual reproduction, both in relation to Darwinian fitness and to long-term phylogenetic effects, are so great that a useful consensus may be some time in coming. Helpful contributions can be expected from a great variety of biological disciplines. I hope that this book stimulates some able theorists to attack the conceptual problems and encourages specialists in diverse taxonomic groups to document those details of life cycles and reproductive natural histories that can provide estimates to replace guesses on crucial quantitative problems.

I did most of the reading and writing during summers of 1970, 1971, and 1972, when I had access to the superb libraries of the University of Toronto and Royal Ontario Museum. I wish to thank the staffs of these institutions for many helpful courtesies. A host of individuals contributed in one way or another to the ideas expressed. Joseph Felsenstein, John Maynard Smith, and Robert L. Trivers worked through the whole manuscript, corrected many errors, and made valuable suggestions. F. James Rohlf and John R. G. Turner helped with conceptual and organizational problems of Chapter 5. Other assistance is acknowledged in the text.

v

PREFACE

I am grateful to the Academic Press for permission to reprint Figures 4 and 5, to Dag Møller and the Fiskeridirektoratets Havforskningsinstitutt for Figure 10, to the Society for the Study of Evolution for Figure 11, to James F. Crow, Motoo Kimura, and the University of Chicago Press for Figure 14. Original Figures 3, 15, and parts of 13 were drawn by Sigurður Gunnarsson of Reykjavik, Iceland, and the rest by Karen Hendrickson of Stony Brook, New York.

Contents

AND CHAPTER SUMMARIES

most of the space. Genetically diverse are more likely than uniform progenies to include the fittest few. If this advantage more than balances the cost of meiosis, asexual reproduction will disappear.

Reasons are given for believing that sexual reproduction will augment numbers at the top end of the fitness distribution when adult movements are limited, fecundity is high, and young are widely dispersed, a common pattern in marine organisms. With sufficiently intense selection, this advantage may balance recombinational load and the cost of meiosis. Evolutionary equilibrium occurs with exclusively sexual reproduction.

Models in Chapters 4 and 5 demand that selection in high-fecundity populations is more intense than is generally realized. Arguments in support of this proposition and implications for its acceptance are discussed.

Published information on adaptive performance, on competitive relations among developing young, and on gene-frequency gradients in relation to dispersal support the proposition of enormous viability variation among genotypes in high-fecundity populations.

Data on higher plants and on fishes indicate enormous variation in fertility. This variation must be partly genetic and contributes to variation in fitness.

Processes envisioned in Chapters 4 and 5 result in
exclusively sexual populations. Their phylogenetic
descendants may lack preadaptations for secondary
acquisition of asexual reproduction, even where it
would be adaptive. Parthenogenesis, where physiologically feasible, rapidly replaces sexual reproduction
in low-fecundity organisms.

Explanations are suggested for some of the phenomena of comparative sexuality: anisogamy, hermaphroditism, selfing, parthenogenesis. Some of their
relations with previously proposed models are
discussed.

Differences between male and female reproductive
behavior and physiology follow from egg-sperm contrast in size. Other factors such as internal fertilization and territoriality predispose a species towards
certain evolutionary changes in the roles of the sexes.
Much of courtship and family life is interpretable as
resulting from partly conflicting male and female
strategies.

Recent literature on this topic contains a diversity of
opinion, but most of the work is based on the assumption that ability to incorporate favorable mutations commonly limits the rate of evolution, and that
recombination must affect this ability. It is proposed instead that potential rates of gene substitution
are always greater than actual, and that recombination is significant mainly for maintaining genotypic
versatility in unpredictable environments.

CONTENTS

Sex and
Evolution

CHAPTER ONE

An Important Question, It's Easy Answer, and the Consequent Paradox

Organisms adapt to stresses by physiological or behavioral responses, as in man's reaction to cold by peripheral vaso-constriction, increased metabolism, and warmth-seeking behavior. That these are normal responses, shown only by living individuals, is evidence that they are adaptations to cold. Logically this evidence is independent of any understanding of insulation or metabolic heat production. A similar approach can be made to the question "What use is sex?" recently posed by Maynard Smith (1971A). If it can be shown, for a variety of organisms that can reproduce both asexually and sexually, that they usually reproduce asexually, but use sexual reproduction in special situation s, the answer arises from inspection. Sex is an adaptation to special situation s. This is a valid conclusion regardless of possible ignorance as to why it should be adaptive in relation to s.

So the answer to the question "What use is sex?" is that proposed by Bonner (1958): [sex is a parental adaptation to the likelihood of the offspring having to face changed or uncertain conditions.] For instance, if the life cycle (zygote to zygote) includes several asexual and one sexual generation, the sexual reproduction will occur where ecological differences will be greatest between two successive generations. Where both asexual and sexual reproduction can occur simultaneously, the asexual offspring will develop immediately and near the parent, but dormant, widely dispersed propagules will be produced sexually.

Bonner noted a number of examples of conformity to this pattern, and there can scarcely be any doubt on its validity. [Many higher plants produce vegetatively a group of genetically identical individuals and at the same time sexually reproduce seeds that are widely dispersed and dormant. Many parasites reproduce asexually in or on a host but sexually when the young are to be dispersed to different hosts. Free-living lower plants and animals that regularly reproduce asexually and only rarely sexually always make the sexual mode a response to changed conditions or to stimuli predictive of changed conditions.]

COMPARISON OF ASEXUAL AND SEXUAL
REPRODUCTION

Asexual progenies are mitotically standardized, and sexual ones meiotically diversified. Besides this essential difference there are less basic but rather consistent contrasts in the natural history of the two modes of reproduction (Table 1).

TABLE 1. Expected differences between asexually and sexually produced offspring

asexual (mitotically standardized) offspring	sexual (meiotically diversified) offspring
large initial size	small
produced continuously	seasonally limited
develop close to parent	widely dispersed
develop immediately	dormant
develop directly to adult stage	develop through a series of diverse embryos and larvae
environment and optimum genotype predictable from those of parent	environment and optimum genotype unpredictable
low mortality rate	high mortality rate
natural selection mild	natural selection intense

The tabulated differences should be sought in those organisms that reproduce both ways. They relate, not to absolute

4

values, but to the direction of a difference where there is one. There are doubtless many organisms in which asexual and sexual propagules are much the same size and about equal in dispersal, but if there is a major difference, the sexual propagule will be smaller, more widely dispersed, and so on. Minor exceptions occur only when morphologically similar structures (eggs or seeds) are used in both asexual and sexual reproduction.

The seasonal limitation on sexual reproduction would result from restriction to that part of the year when predictability of conditions from parental to offspring generation is at its annual minimum.

Localized development of asexual propagules follows mechanically from their large size and immobility in plants. Seeds are small in comparison, and show a wealth of special adaptations for securing wide dispersal. That vegetative structures readily evolve means of dispersal is clear from the many examples of such structures that are not themselves capable of further development, but form a vehicle for seed transport. The polyp-medusa cycle of coelenterates may seem to be an exception, because budded-off medusae may account for more dispersal than sexually produced planulae. From the polyp's point of view it conforms to the rule. The budding-off of medusae is its initiation of sexual reproduction, and the next polyp generation will be genetically diverse and develop far away as a result of the dispersal of medusae and planulae.

Initial dormancy in sexual propagules can be an adaptation to expected unfavorable conditions, as in autumn-shed seeds where winters are severe. No such function can be attributed to dormancy in seeds shed early in the growing season, but dormancy will always effect dispersal through time. This is especially true of seeds that are widely variable in duration of dormancy. A single plant may produce seeds that vary in dormancy from months to many years, and long- and short-dormancy seeds can be visibly different. Epling, Lewis, and

5

Ball (1960) and Harper (1965A and B) review these phenomena, and a theoretical treatment of adaptive dormancy is provided by Cohen (1968). His reasoning assumes population-wide optimization. I believe it to be equally applicable, and more biologically relevant, to progeny optimization.

Differences in mortality rate must follow from differences in size, dispersal, and dormancy. Large organisms generally have lower mortality rates than small, and large size indicates greater resources. Long-distance dispersal must be especially hazardous when favorable habitats are small and isolated. Long dormancy increases the time during which death from accident, disease, or predation can occur.

The accuracy with which offspring environment can be predicted from parental environment is obviously related to dispersal through time and space. I interpret the direct development of asexual offspring and the indirect development of the sexual as part of the same pattern. An adult organism developing in a certain place is living proof that the habitat is favorable, perhaps extraordinarily so, for an adult of that genotype, and that in the past it was favorable for earlier stages of development. It is not especially likely that this habitat is now a good one for an embryo or larva or seedling, which will have requirements and challenges different from those of the adult. Asexual production of offspring in the immediate vicinity of the parent seems to be adaptive only when an offspring quickly becomes adult in its ecological requirements. If it will be ecologically different as a result of being ontogenetically different, genotypic diversification is the normal strategy.

The comparison of intensities of selection of asexual and sexual reproduction requires some elaboration. If one compares selection within progenies the answer is tautologically evident. There can be no natural selection in the genetically uniform progeny. For a comparison of selection between two asexual progenies with that between two sexual ones, the difference is the reverse of that claimed. The intended comparison is the hypothetical one of intensity of environmental scrutiny of genotype. Natural selection of widely dispersed seeds and seedlings of strawberry would be more intense than that of a sample of runner-propagated individuals of the same genetic diversity.

THE PARADOX OF SEXUALITY

I know of no observations clearly counter to those summarized above. Henceforth I assume that the association between sexual reproduction and changed conditions, a relationship recognized by many and discussed in detail by Bonner, is adequately supported, even though the nature of the relationship and of what is implied by changed conditions is not yet clearly specified. So I assume that the answer to the question "What use is sex?" is settled, but in a way different from Maynard Smith's conclusions. He found little support for the idea of sex as an immediately adaptive feature of reproduction and argued for a form of the traditional view (elaborated by H. J. Muller, therefore *Mullerian,* see Chapter 12) that emphasizes long-term effects on gene substitution. My reasoning is much less rigorous than Maynard Smith's thorough mathematical treatment, but I think that I have the more reliable conclusion. For answering questions on function in biology, comparative evidence is more reliable than mathematical reasoning.

Exact reasoning from plausible assumptions on the natural selection of variations in processes of sexual reproduction is obviously necessary. The conclusion that sex is an adaptation to changed conditions is not very satisfying until we understand to some degree why it is adaptive to changed conditions. Fortunately it is easier to reason from a set of premises to a valid conclusion if you know the conclusion in advance. Having learned, from comparative evidence, what sex is

7

adaptive to, should make it easier to show, by deductive analysis, why it is adaptive.

This book will be largely devoted to answering this secondary question. The task would seem immensely difficult, as Maynard Smith (1971A, 1971B) has clearly recognized, because we can immediately see an enormous disadvantage in sexual reproduction. Just how enormous is a complicated question. I assume for the moment that an evaluation can be made by comparing two kinds of individuals. One is parthenogenetic and produces diploid eggs of parental genotype. The other produces genetically diverse haploid eggs that require fertilization. The relative success of these two individuals will specify the relative fitness of asexual and sexual reproduction. On this basis the arguments below show that the parthenogenetic individual has twice the fitness of the sexual. This immediate advantage of asexual reproduction is generally conceded by those who have seriously concerned themselves with the problem. If the reader has any doubts on this matter as discussed in the paragraphs below, he can find a more exact treatment in Maynard Smith's (1971B) discussion.

Consider a population with two female genotypes. At a certain stage in the life history, genotype A_1A_2 produces unreduced eggs that develop into genetic replicas of the mother. Genotype A_3A_4 produces reduced eggs that must be fertilized. An A-allele from a sperm can be called A_m. Thus an offspring of the sexual parent is either A_3A_m or A_4A_m. It has one, not both of the maternal genes. All the offspring of the parthenogenetic parent have the full maternal genotype A_1A_2. Unless something causes a difference in the numbers of offspring, the asexual parent has double the genetic representation of the sexual in the offspring generation. It is the same story in the next generation. Both genes of the asexual progenitor will be present in each of the grandchildren, but only half of the descendants of the sexual individual will have either A_3 or A_4, which are therefore down to a quarter of their original fre-

8

quency. Each "sex gene" suffers a 50% hazard per generation, relative to asexual alternatives. Purists who object to the implication that an asexual clone is a part of a Mendelian population can endow the genes for parthenogenesis with less than perfect penetrance.

A gene that causes sexual reproduction only in certain genetic and environmental circumstances will suffer the 50% loss only infrequently. The character still has the 50% disadvantage, even though the effect may be infrequent or diffused over many loci. Also, any advantage we might attribute to the character will likewise be infrequent when penetrance is low. The argument so far applies only to females. The impossibility of parthenogenetic development of sperm makes it inapplicable to males, unless a male can reproduce vegetatively, as discussed below. The reasoning assumes the possibility of producing diploid eggs, so it does not apply to haploid organisms.

An important assumption is that diploid and haploid eggs are equally expensive. This can only be approximately true, but the inaccuracy would usually be minor. There is only half as much DNA in a haploid nucleus, but there is no reason to believe that DNA is especially difficult to produce, compared to RNA or proteins. Even if DNA synthesis made extraordinary nutritional demands, meiosis would not achieve any saving where polar bodies are shed with the eggs. In species that resorb polar bodies the DNA saving would still be slight, because nuclear DNA is only a small fraction of that in the egg. Even the minute egg of the sea urchin has much more DNA in yolk and mitochondria than in the nucleus (Brachet and Malpois, 1971). Reduced and unreduced eggs may be nearly identical physically and chemically. The important difference lies in genetic information content.

The concept of *cost of meiosis* is clearest when the comparison is between sexual and parthenogenetic egg production, but is equally valid in relation to sexual egg production versus

vegetative reproduction. Consider an asexual species that reproduces by both parthenogenesis and budding. Eggs and buds
may differ in many ways, and a given mass of one may represent a different expense to the parent from the same mass
of the other. Yet it must be true, at evolutionary equilibrium,
that average return per investment is the same for the two
processes. If it were more expensive to produce an offspring
by one process, the species would evolve a lower level of investment in that process until the difference disappeared. It would
disappear if the cost-benefit relation were frequency dependent so that the originally less efficient process became more
efficient as it became less frequent. Otherwise the less effective
mode would simply be lost, and the species would be left with
either budding or parthenogenesis, not both.

Suppose frequency dependent selection operates and we
have a species in which the two processes are at equilibrium.
The profit rate, in numbers of offspring and in genes transmitted, is the same for parthenogenesis and budding. Now
create a second species, identical in every way, except that
its eggs are haploid and require fertilization. Even if the necessary sperm is always provided, an egg of this second species
is only half as effective in gene transmission as that of the
first. It must therefore be only half as effective as its own or
the other species' budding. We would now expect egg production to be rapidly lost in the second species. Only if the frequency-dependent effect were so powerful that the effectiveness of the eggs were doubled at some lower frequency would
reproduction by haploid eggs be retained. The eggs would
then be equally as effective as buds in gene transmission, and
twice as effective in producing offspring. This relationship
must be true of all species that retain both sexual and asexual
reproduction.

The 50% cost of meiosis applies to outcrossed but not to
self-fertilizing hermaphrodites. Hermaphroditism and isogamy
are discussed further in Chapter 10. If males assist females

10

in raising the young, the cost of meiosis is reduced (Maynard Smith, 1971A), unless it is possible for a female to get a male to help her raise parthenogenetic young.

The primary task for anyone wishing to show favorable selection of sex is to find a previously unsuspected 50% advantage to balance the 50% cost of meiosis. Anyone familiar with accepted evolutionary thought will realize what an unlikely sort of quest this is. We know that a net selective disadvantage of 1% would cause a gene to be lost rapidly in most populations, and sex has a known disadvantage of 50%. The problem has been examined by some of the most distinguished of evolutionary theorists, but they have either failed to find any reproductive advantage in sexual reproduction, or have merely showed the formal possibility of weak advantages that would probably not be adequate to balance even modest recombinational load. Nothing remotely approaching an advantage that could balance the cost of meiosis has been suggested. The impossibility of sex being an immediate reproductive adaptation in higher organisms would seem to be as firmly established a conclusion as can be found in current evolutionary thought.

Yet this conclusion must surely be wrong. All around us are plant and animal populations with both asexual and sexual reproduction. Can we seriously consider that the quantitative apportionment of resources to these two processes is not subject to Darwinian selection? Do aphids, coelenterates, and various higher plants seem to be evolving a reduced frequency of sexual reproduction? Only consistently negative answers are possible for these questions. There is no escaping the conclusion that these life cycles must be close to evolutionary equilibrium. The observed incidence of asexual and sexual reproduction must represent for these forms the currently adaptive optimum maintained by selection. In these populations there can be no net disadvantage to sexual reproduction.

So that which must surely be false, by the method of deductive analysis, must as surely be true by comparative evidence, and vice versa. My approach to this paradox is to accept the comparative evidence and try to revise the analysis so as to make the two compatible, although my revision requires what some might consider rather drastic modifications of current evolutionary thought.

PREVIOUS SUGGESTIONS ON SEX AS A REPRODUCTIVE ADAPTATION

A belief that sex may be adaptive by diversifying offspring potentiality in relation to uncertain conditions dates at least from Weismann (1889), but critical discussions are few. Clausen (1954) proposed a model of evolutionary equilibrium between asexual and sexual reproduction, but only in reference to rare sexuality in normally apomictic hybrid plants. His reasoning was not quantitative and did not recognize the cost of meiosis. In a critique of group selection I proposed (Williams, 1966A) that all long-term advantages that had been proposed for sexuality were valid as immediate adaptations for maximizing progeny success. I pointed out that the association of sex and changed conditions in heterogonic (alternating asexual and sexual) life cycles implied adaptation to uncertainty. Ghiselin (1969) also suggested progeny success as the important consideration and criticized the Mullerian theory for ". . . failing to account for the increase of genes for sexuality within the population." Maynard Smith (1971A) analyzed the possibility of progeny diversity as a reproductive adaptation and found it unlikely to produce any advantage. This was with both a general model, without any special life-history features, and a form of the Aphid-Rotifer Model (Chapter 2). Williams and Mitton (1973) reexamined the Aphid-Rotifer Model and concluded that the special life-history features of the model organisms would result in a frequency-dependent advantage in sexual reproduction such that

12

evolutionary equilibrium would occur with several to many asexual generations per life cycle. They also introduced the Elm-Oyster Model (Chapter 4).

Emlen (1973:54–56) presented a graphic model of the favorable selection of sexuality. Frequency distribution of offspring on a scale of fitness dosage is shown as having the same mean for both an asexual and a sexual progeny, but the sexual has the broader distribution. If selection is so intense that only those at the upper end of the scale have an appreciable chance of survival and reproduction, there may be more successes in the sexual progeny, which would therefore have the higher mean fitness. This would depend on a nonlinear relation between position on the dosage scale and fitness. Emlen uses a threshold relation, with only those above a high dosage being "fit," and all those below being "unfit."

This is justified in a brief introduction such as Emlen's, but sharp fitness thresholds are unlikely in nature. I prefer a more complex representation (Figure 1), where a fitness gradient (shown as density of stippling) replaces Emlen's threshold. I have also lowered mean dosage in the sexual progeny to indicate recombinational load.

A qualification is necessary. It is not enough that a sexual progeny be more variable than the asexual. What is required is that a sexually produced individual have a wider probability distribution of fitness, and this need not always be true. For instance, if the parental generation is constituted from extreme types, which would produce an intermediate F_1, the offspring group would be less variable if produced sexually. The relationships among population and progeny fitness frequencies and individual fitness probabilities are discussed further in Chapter 5.

The models proposed here (Chapters 2 to 5) all assure a greater variance in fitness dosage among sexually produced individuals, and are therefore all examples of Emlen's more general model as modified here. These assurances come from

1 3

special life-history features, such as intesnse competition within progenies, and special breeding structures. I believe that additional models may be valid, but that all will require special life-history features. Maynard Smith's (1971A) analysis

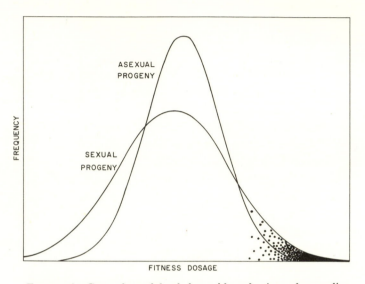

FIGURE 1. General model of favorable selection of sexuality as a result of increased fitness variation. The sexual progeny has the lower average dosage of fitness, because of recombinational load, but extends to higher values because of its greater variability. Density of stippling indicates fitness.

convinces me of the unlikelihood of anyone ever finding a sufficiently powerful advantage in sexual reproduction with broadly applicable models that use only such general properties as mutation rates, population sizes, selection coefficients, etc.

CHAPTER TWO

The Aphid-Rotifer Model

Suppose you were offered this choice in a lottery: either you could have several different tickets, or you could have the same number of copies of the same ticket. Obviously you would elect to have several different tickets. The theory presented in this chapter claims that sexually produced offspring may be analogous to lottery tickets, and those asexually produced analogous to redundant copies of the same ticket. If and when the analogy is valid, sexual reproduction is more likely to produce winners.

This chapter will attempt to show, for heterogonic life cycles, that genetic diversification of progeny may carry at least a two-fold advantage and thereby justify the cost of meiosis. It enlarges on the Aphid-Rotifer Model proposed by Williams and Mitton (1973) but, needless to say, leaves many interesting questions untouched. My purpose is merely to argue the likelihood of an advantage in sexual reproduction that is frequency dependent in a way that may explain the relative frequencies of sexual and asexual reproduction in life cycles such as those of aphids, rotifers, and many parasites. Subsequent chapters will deal with other kinds of life cycles.

THE MODEL

Consider a population with the following life history. It lives in small, discontinuous habitats, such as a host organism or a woodland pool. It is diploid and produces a number of generations a year asexually so that a single individual, on reaching a suitable habitat, can form a clone that saturates the limited environment. Once a year a generation of widely dispersed diploid propagules is produced. The Model relates to the relative advantages of asexual and sexual reproduction for the production of these widely dispersed propagules.

Aphids that form clones on herbaceous hosts in seasonal climates would be an obvious example. Rotifers, turbellarians, and other small invertebrates of temporary waters would be others. Some protists have the appropriate ecological life history and lose genetic material in the formation of macrogametes, e.g. the diatoms (Fritsch, 1935; Prescott, 1968) and perhaps some sporozoans (Kudo, 1966). The model may be applicable to them, but it is irrelevant to the many haploid isogametic protists.

In the production of widely dispersed propagules let genotype a produce them asexually, so that all have the parental genotype a. Genotype b produces them sexually, so that they have genotypes c, d, e, etc. An average of λ propagules from each parental genotype establish themselves in accessible new habitats, and each then proliferates clonally until the habitat is saturated. Thereafter the clone with the locally best genotype will continue to increase until it is the exclusive or dominant clone.

Lineages established in a habitat may be m of genotype a, from the asexual parent, and one each of genotypes i, j, k (a total of n) from the sexual parent. There is only one chance in $1 + n$ that the best adapted clone will be a. It is initially the most numerous, its starting numbers being expected to equal all the others combined, but after a long period of competition in a saturated environment, the genotypic proportions may be decided more by minor differences in physiology than by major inequalities of initial abundance. Being n times as likely to produce the one fittest genotype may ultimately more than compensate for the twofold cost of meiosis.

The envisioned advantage of sexual reproduction is that it gives one's genes representation in a variety of clones, thereby

1 6

increasing the chance of transmitting them to winning clones. There is a close formal analogy with the lottery tickets. An important assumption is that the propagules from one parent (or parental clone) may have access to only a few new habitats, but that more than one, perhaps many, may reach a habitat that is accessible. From the habitat's standpoint, colonists are a limited and very nonrandom sample of the entire parental population. Another important assumption is that clonal descendants of multiple propagules compete with each other for a level of success that one propagule could achieve by itself. Increasing the number of genotype-*a* zygotes reaching the water in a bromeliad may have no more influence on the final size of clone *a* than xeroxing one's lottery ticket would have on one's winnings.

The envisioned processes imply enormous differences in genotypic selection coefficients in one zygote-to-zygote cycle (hereafter *life cycle,* as opposed to *generation,* which may be either asexual or sexual). The more or less complete replacement of one clone by another in a several-month growing season is ecologically realistic, but it corresponds to a manifold difference in selection coefficients per life cycle. The 50% disadvantage of meiosis may be merely one of a large number of strong selective effects and need not be decisive.

It might be objected that asexual propagules, in retaining a genotype that proved its fitness in a previous clonal competition, has a better than average chance of winning the next competition. Any such between-habitat heritability of fitness would reduce the proposed advantage of sexuality. Gill's (1972) experiments show that minor differences in conditions, comparable perhaps to differences between successive habitats, can influence outcome of competition between strains of Paramecium. Between-habitat heritability of fitness would apparently be low in these organisms.

Relative fitness of genotypes must also change during a period of clonal competition, as a result of increased crowding and perhaps seasonal changes. The fitnesses assumed in the model must be averages over the period of competition, and it is assumed that average differences are great enough to cause considerable change in relative abundance in a few generations. There is a formal possibility of clones coexisting in numerical equilibrium, but the conditions are stringent (B. Levin, 1971; Stewart and Levin, 1973). If the principle of competitive exclusion is ever valid, it should be valid for clones of the same population competing within a circumscribed habitat.

The various plausible ways in which the advantage of genetically diverse progeny can be reduced do not detract from the general logic of the model. They can be compensated merely by prolonging the period of clonal competition, or by increasing the number of propagules reaching each new habitat, as indicated in the next section.

FREQUENCY OF SEXUAL REPRODUCTION IN EVOLUTIONARY EQUILIBRIUM

The longer the periods of clonal competition, the more will differences in genotype result in differences in abundance. So the more generations per life cycle, the greater the advantage in the mixed strategy of sexuality. If this advantage is more than twofold, sex has a net selective advantage and should increase in frequency. This increase will reduce the length of the life cycle and thereby reduce the advantage of sexuality. Evolutionary equilibrium occurs when the mixed-strategy advantage just barely pays for meiosis.

Williams and Mitton (1973) simulated this relationship on a computer. They calculated relative numbers of asexual descendants of two sets of propagules in a circumscribed habitat. One set from one parent (or parental clone) was asexually produced and all had the same fitness e^{x_0}. The other set was sexually produced by a pair of parents or parental clones and had various fitnesses e^{x_i}. Colonized habitats were equally accessible to both sets of propagules, so that $\bar{m}$ (average number of established asexual propagules) equaled $\bar{n}$ (average number of established sexual propagules). Both m and n were chosen from the same Poisson distribution (mean = λ). Fitnesses e^x were obtained from a table of random normal numbers with $\bar{x} = 0$ and $\sigma_x = 0.1$. The standard deviation of 0.1 was considered realistic when asexual reproduction is by binary fission, but higher values were considered more likely with prolific parthenogenesis.

After t generations in the colonized habitat, the descendants of the sexual propagules would make up

$$\frac{\sum_{i=1}^{n} e^{tx_i}}{\sum_{i=1}^{n} e^{tx_i} + me^{tx_0}}$$

of the total.* Averages for a number of values of t and λ (Table 2) showed that this ratio reaches the critical value of $2/3$ after from 12 to more than 100 generations, with shorter times for

---

* Professor Robert M. May has provided the following commentary on this ratio.

If x be chosen randomly from a normal distribution with mean x_0 and variance σ^2, the average value of the finess function e^{tx} is

$$(2\pi\sigma^2)^{-\frac{1}{2}} \int \exp\left[tx - (x - x_0)^2/2\sigma^2\right] dx = e^{tx_0} e^{t^2\sigma^2/2}$$

Thus, particularly if we are dealing with a relatively large mean number of propagules (large λ) so that statistical fluctuations are smoothed out, the individual selective advantage of a sexual over an asexual propagule is given by the factor $\exp(\frac{1}{2}t^2\sigma^2)$. Therefore the sexual form will have a *net* selective advantage once this factor exceeds 2, i.e. once $t^2\sigma^2 > 2/n2$, or $t > (1.18)/\sigma$. Table 2 derives from the specific assumption $\sigma = 0.1$, whence we expect the sexual form to predominate once $t > 12$, which accords with the numerical simulations.

TABLE 2. Descendants of sexual propagules as a proportion of total number of competitors. The heavy lines separate those combinations of propagule number (λ) and number of asexual generations (time) in which asexual (above the line) and sexual (below the line) has a net selective advantage (from Williams and Mitton, 1973).

time	mean propagule number (λ)								
	3	4	6	9	12	16	25	50	100
2	.51	.49	.51	.50	.50	.51	.51	.51	.50
4	.50	.52	.52	.52	.53	.51	.51	.51	.53
7	.53	.54	.55	.54	.56	.55	.53	.54	.55
12	.52	.59	.59	.61	.63	.58	.62	.66	.67
20	.65	.59	.64	.70	.73	.71	.74	.74	.73
30	.61	.69	.72	.77	.75	.79	.80	.81	.85
50	.62	.70	.74	.82	.87	.84	.86	.96	.92
100	.64	.73	.81	.81	.90	.88	.93	.93	.99
∞	.69	.76	.83	.89	.91	.93	.95	.97	.98

larger values of λ. Equilibrium times also decrease in inverse proportionality with σ_x. If $\sigma_x = 1$, only one-tenth as much time is required. The model is merely illustrative, and cannot predict durations of clonal multiplication, because only guesses are possible on values of σ_x and λ, and because other factors that would affect the selection of sexual reproduction in nature are left out of account. These other factors could make the critical value rather different from $2/3$. Use of a single of e^x for each genotype does not mean that each clone continues to increase at the same exponential rate. It merely means that ratios of such exponential functions continue to represent ratios of clonal abundance. Each e^x represents a *mean* relative rate of increase up to time t.

The important properties of the model are not dependent on the form of the distribution of x or on fitness being distributed as e^x, or on asexual and sexual propagules having exactly the same fitness distribution. All that is required is that the distributions be roughly the same, and that within-habitat heritability of fitness is much greater than between-habitat heritability.

20

PARAMETERS OF HETEROGONIC LIFE CYCLES

If stability is achieved after t generations, and if fecundity per generation is F, the potential increase per life cycle is F^t. If selection is more intense with higher fecundity, as will be argued in later chapters, F and t should be inversely related, with equilibrium being reached with fewer generations in more individually prolific organisms. The natural history of heterogonic animals may give some support to the proposition that F^t (potential zygote-to-zygote increase, henceforth ZZI) is rather invariant over a diverse taxonomic array of organisms.

For example, parthenogenetic females of *Daphnia pulex* can produce about 40 young per lifetime, with generation length about three weeks at room temperatures (Frank, et al., 1954). Under average field conditions in a six-month growing season, generation length would probably average more like six weeks. A single zygote at the start of the season could thereby give rise to about ten million by the end ($ZZI = 10^7$). A flatworm that divided by fission once a week for the same period would have about the same ZZI. A protist with fission every day or two would have to have several sexual phases per year to keep its ZZI down to ten million. A careful study of fecundities and numbers of generations per life cycle of a variety of heterogonic animals would be of great interest.

Either the physiological feasibility or selective value of sexual reproduction, or both, may be contingent on environmental cycles. Sexuality for Daphnia may have greatest value at a certain time of year, and day length or other cues may be utilized for synchronizing the developmental processes leading to mictic individuals. An annual cycle would be feasible, but not a slightly shorter or longer cycle. Seasonality of life cycles may be rather resistant to evolutionary change.

In relation to the model, there could still be slight adjustments in the average frequency of sexual reproduction. This would result from change in proportions of mictic individuals.

In a Daphnia culture, some individuals may form winter eggs while others continue to reproduce parthenogenetically, at least for a while, and it is possible for winter eggs to be produced parthenogenetically (Bacci, 1965). So if some genotypes occasionally skip the sexual phase while others indulge, the ZZI will readjust itself after a long-term environmental or evolutionary alteration in its value.

The same readjustment could be made by altering the fecundity per generation. Suppose a population were at equilibrium with an average of 4 parthenogenetic generations per annual cycle and 56 eggs per generation ($ZZI = 10^7$). An environmental or genetic change that allowed time for an average of only 3.9 generations would make annual sexuality of above optimal frequency. Clones with greater than average individual egg production would have nearer the optimum ZZI and would be favorably selected. In the absence of other changes, equilibrium could be restored by increasing fecundity to about 60 eggs.

In $ZZI = F^t$, neither F nor t has an upper limit. The lower limit for F is two (binary fission) and for t is one (no clonal multiplication at all). If evolutionary equilibrium occurs at a certain ZZI, it follows algebraically that there is a value of F at which the requisite ZZI would be obtained with one generation per life cycle. Above a certain level of lifetime female fecundity and with competition within habitats of the sort envisioned in this chapter, we might expect to find exclusively sexual organisms. They would logically form a special case of the Aphid-Rotifer Model, but it is convenient to treat them with a model that is at least superficially different (Chapter 4).

OPTIMAL TIMING OF SEXUAL REPRODUCTION

Suppose there are several periods of multiplication in a life cycle, as in some trematodes. The question arises: which of these reproductive phases should be sexual. An answer can be found in the arguments presented. Sexuality should occur where there is minimal fitness heritability and maximum likelihood of new genotypes being of greatest fitness. In other words, where conditions on the right of Table 1 (page 4) are maximally developed.

Some examples of association between sexual reproduction and changed conditions were given in Chapter 1. A long list of animal parasites that conform to the pattern could be given. Reproduction within hosts is asexual; production of propagules to colonize new hosts is sexual. Where there is more than one obligate host, the final host (where sexual reproduction takes place) will be the most mobile and disperse the parasite most widely. Thus the frequent trematode pattern of a snail for intermediate host, and vertebrate (or squid) for final. Rather than belabor these instances of conformity to expectation, I will here mention some doubtful or seemingly exceptional cases. They are taken from general works on parasitology (Baer, 1952; Cable, 1971; Cheng, 1964).

There is a sporozoan in which a turtle is the secondary and a leech the final host. Neither would obviously disperse the parasite more widely nor be harder to find than the other. For the malarial parasite it is not at all clear whether mosquito or vertebrate is more widely dispersed, or which transfer between hosts should be subject to greatest selection. Of course this parasite conforms to expectations to the extent that reproduction within hosts is asexual. Only in the production of propagules for transfer from mosquito to vertebrate is reproduction sexual.

Round worms often lack asexual reproduction and are seldom relevant to the discussion. There are a few in which sexual reproduction alternates with a single parthenogenetic generation. I cannot conceive of this as an evolutionary equilibrium. I suggest that an ancestor had many parthenogenetic generations per life cycle. The clonal period may then have been restricted by an environmental change, and the population may have lacked preadaptations for reestablishing the equilibrium. In one roundworm with a single parthenogenetic generation the reproductive patterns are the opposite of those predicted. The infective stage is a fertilized female, which releases genetically diverse young within a single maggot. These young produce, parthenogenetically, individuals that will leave the maggot and copulate in the soil.

The association of sexual reproduction with changed conditions is also understandable as an adaptation for free-living organisms, and many examples could be given from general works on lower plants and animals. I will here note some relevant facts in relation to some better known examples.

Winter eggs of Daphnia are not necessarily produced in anticipation of winter. Drought, high temperature, unfavorable food, may all stimulate their production. Whatever the unfavorable condition, the genetically diverse winter eggs lie dormant until favorable circumstances return. They may return in the fall, or perhaps not until the following spring. Relations among temperature, rainfall, and light may be rather different in successive springs. Flooding and other factors may disperse dormant eggs to different localities. There can be no doubt that it is the dormant eggs that will produce the generation with conditions least predictable, and these are the eggs usually produced sexually.

The sexual phases of Daphnia and other small invertebrates of large lakes may be infrequent or absent. This is understandable on the basis of the greater stability and predictability of conditions in large, permanent bodies of water. Unfortunately for this line of reasoning, it is the clonal phases that are missing among marine representatives. I can only speculate that in a stable marine environment, any small motile invertebrate that can reproduce asexually at all will have that process completely replace the sexual. Exclusively asexual populations may then have a high probability of extinction (Chapter 13).

Some recent work on the population genetics of rotifers supports the model. King (1972) showed that there are great seasonal changes in clonal composition in wild populations. One of his proposed explanations (called "complete genetic discontinuity") conforms to the model proposed here. The other does not.

This chapter, and the three to follow, concentrate on finding a major advantage of sex, of the sort indicated in Figure 1 (page 14), to balance the major cost of meiosis. Admittedly this is too limited a view and other factors will warrant consideration. Beardmore (1963), Ghiselin (1969), and Maynard Smith (1971A, 1971B) have discussed some of the additional selective forces that must influence the relative advantages of sexual and asexual reproduction. Probably the most important additional disadvantage of sex is that it generates recombinational load. Obligate outcrossing would also handicap a fugitive species in which colonization of temporary habitats may depend on single propagules. Minor advantages to sexual reproduction can be seen in possible escape from vegetatively transmitted pathogens, reduced susceptibility to contagion, and reduced competition (as shown by Beardmore) in genetically diverse progenies.

CHAPTER THREE

The Strawberry-Coral Model

Chapter 2 dealt with organisms that may move freely but only within limited, isolated habitats during clonal multiplication. This chapter deals with sessile organisms that multiply vegetatively in continuous habitats. Their widely dispersed propagules are sexually produced. The strawberry, with reproduction by runners in addition to sexual seed production, is a good example. A coral, with vegetative proliferation of largely independent polyps and sexual production of planulae larvae, is another.

THE MODEL

When a young strawberry or coral is successfully established it starts making adjacent copies of itself by asexual reproduction. This vegetative increase can theoretically continue without limit. Actually the spread continues only so long as peripheral individuals continue to find favorable conditions. Eventually the clone meets tolerance limits in all directions and the spread ceases. Shifting thresholds in environmental gradients thereby substitute for physical limits of habitat space assumed for the Aphid-Rotifer Model. A strawberry or coral clone is not limited to a season or the lifetime of a host, but sooner or later it must come to an end. Grossly adverse physical changes may occur, or a better adapted clone, for the area occupied, may become established and crowd it out.

A graphic model, with clones assumed to be in short-range equilibrium between growth and attrition, makes the matter easier to discuss (Figure 2). Blackened circles represent members of one typical clone, now made up of 15 individuals. Each one may be short lived, but when one dies another will take its place so that number, biomass, and area of clonal occupancy will vary only slightly over significant periods of time. A few of the many likely environmental gradients are represented: grazing pressure, light, competition with other clones,

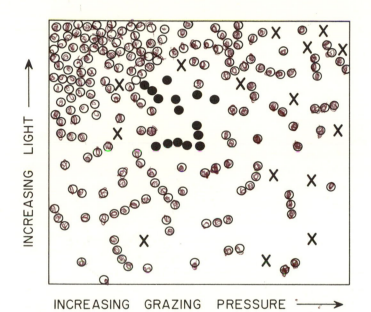

FIGURE 2. Strawberry-Coral Model. Dark circles are members of one clone, open circles members of other clones, X's members of a competing species.

and with members of another species. In its limited region of this complex field our clone is, for the moment, better adapted than any other. If all other clones were removed, it would spread further, especially towards the upper left, where it is now excluded by competing clones, but any one clone should have a smaller area of occupancy than that held by all currently established clones.

Area of occupancy and biomass are determined by the equilibrium between capture of resources by feeding (coral) or photosynthesis (strawberry and some corals) and their use in maintenance and replacement of losses to grazers and parasites. An additional demand, for instance seed production, would require a reduction in resources for maintenance and vegetative growth and ultimately a smaller biomass and area of occupancy. If k is the maximum number or biomass that could conceivably be maintained, the use of a certain proportion of available resources for sexual reproduction would reduce the clone size to pk $(0 < p < 1)$. Sexual reproduction would therefore mean a lower life expectancy, because the smaller the territory, the more likely it would be for all of it to become uninhabitable or lost to a competing clone.

On the other hand, it is clear that any gene will disappear in a geologically short time if it can not untie its fortunes from those of an inevitably mortal clone. Evolutionary success will require vagile propagules capable of establishing themselves beyond clonal boundaries. Since these boundaries define the region of adequacy of the clone's genotype, it is necessary that these propagules be sexually produced. A genetically diverse progeny would have a finite probability of including individuals capable of establishing themselves elsewhere in the environmental gradients and may be worth the 50% cost of meiosis.

The Model may be applicable to a large part of the Earth's vegetation and to much of the marine fauna. Many sponges, sessile polychaetes, bryozoans, protochordates, and other marine animals start life as widely dispersed planktonic young that colonize suitable surfaces. Once established an individual produces a colony that spreads in all directions as long as it can. When two such colonies meet, one will grow at the expense of the other, even if the physical limits of the colonies become obscured in the process. The great majority of initiated clones will be crowded out or overgrown or perish from other causes. Mature colonies will often live in a condition of maximal crowding, with each able to persist for a long time in its limited realm, but unable to spread further along the environmental gradients in which it is found.

SELECTION IN A POPULATION OF
LOCALIZED CLONES

A more detailed look at known or probable processes in the ecology and genetics of such organisms as strawberries and corals adds plausibility to the model. The situation depicted in Figure 2, in any persistent community, may be the outcome of many years of selection for locally fittest genotypes. Those actually present can only be a minute fraction of those that have participated in the contest. The competitive exclusion principle should apply in extreme form. Even a minute advantage for one clone can assure at least the partial elimination of a competitor. Every year enormous numbers of new genotypes enter the field against the established clones, each of which has already demonstrated extraordinarily high fitness for its region of occupancy. A newcomer only has a minute chance of establishing itself where fate locates it in the environmental gradients. Occasionally one succeeds and crowds out one or more older clones, wholly or partly. Thereafter the prevailing clones represent an even more select array.

The strategic value of sex in a population of localized clones becomes clear if fitness is analyzed into a general and a local component. A given genotype tested in a representative sample of natural conditions would be found to have a certain average fitness, but its fitness would vary with position in the habitat. Habitat space could be represented as an adaptive landscape of hills and valleys that would indicate the fitness of every position for the genotype in question. Variation in this local fitness, or at least that part of it that is unrelated to general fitness, would be an independent additional contribution to variation in fitness. The fitness of a given genotype in a given position would be e^{x+y}, where x is the general and y the local component.

With selection of the intensity envisioned, success depends on having both x and y abnormally large. To be in the top millionth of the distribution of general fitness may be of no avail to a propagule that lands at a point where its local fitness is below the median value. It may not be as fit as an individual that is merely in the top thousandth of the general fitness distribution and also in the top thousandth of local fitness. Any persistent clone shows not only that it has an extremely fit genotype in general, but also that it was extremely fortunate in settling in a part of the habitat that is unusually well suited to its genotype. Figure 3 shows the envisioned relationship between fitness and position in a cross-section of such a community as that represented in Figure 2. Clones *A, B,* and *C* currently occupy that part of the gradient represented. All other genotypes have disappeared or failed to get established, either because of low general fitness (for example, *E*), low

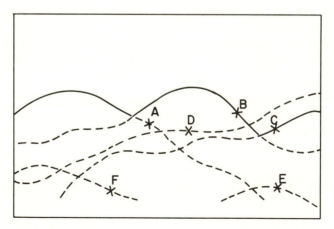

FIGURE 3. Fitness of clones as a function of environmental gradient in a section through part of a habitat space. Solid lines indicate regions of actual occupancy. Broken lines show regions that the genotype would occupy if competing clones were absent. *X*'s mark positions of initial settling of clones *A-F.*

local fitness at position of settling (D), or both (F). A is assumed to have settled and spread to the left before B arrived and usurped the region of A's initial establishment. In nature, the overwhelming majority of genotypes would have general and local fitness like that of F; a few would have high general, but mediocre local fitness, or vice versa, like D or E, and only a minute fraction would be so high in both as to actually win part of the habitat.

Consider now the long-term challenge faced by clone B. It has the fittest genotype, for its limited region, that has ever had access to that region, but if large numbers of new genotypes are being introduced, one may someday prove to be even fitter. That event or an environmental change may cause clonal extinction. To achieve genetic survival beyond its finite clonal survival, B must get its genes represented in other clones. It must disperse genetically variable propagules to regions beyond its current holdings. An important assumption follows from the clone's circumscribed distribution. The inability of B to spread into regions occupied by A and C demonstrates the inadequacy of its genotype in adjacent regions. Somewhere, perhaps, is another locality where B would be fittest, but within the readily accessible region it is likely that conditions grow worse in all directions. Only by genetically diversifying the widespread propagules is there any hope of producing any with sufficiently high local fitness.

VEGETATIVE AND SEXUAL REPRODUCTION IN EVOLUTIONARY EQUILIBRIUM

All organisms are expected to apportion resources according to an optimum compromise between the competing needs of growth, maintenance, and reproduction. Strawberries and corals must also achieve an optimum in apportioning resources between two modes of reproduction. To balance the cost of meiosis, a given investment in sexual reproduction would have to have twice the profit in offspring of the same investment in asexual reproduction.

31

Asexual reproduction clearly shows a profit over cost as long as the clone is spreading. When the spread has stopped, investment in asexual reproduction is still necessary to maintain the size of the clone, but the only "profit" is yield to grazers and parasites. Maximizing clonal size is important to sexual reproduction; the larger the clone, the more zygotes it can produce. At the same time, investment in sexual processes diverts resources from clonal maintenance and must cause a smaller clonal size. The balance between rates of asexual and sexual reproduction is assured by both processes reaching a point of diminishing return with increasing investment.

An individual well within the region held by the clone would have an especially low profit on investment in asexual reproduction. Increasing the density of genetically identical individuals in its immediate vicinity would intensify competition within the clone. Investment in widely dispersed (and therefore sexual) propagules would avoid this waste. So centrally located individuals would be expected to invest more in sexual reproduction than would peripheral ones. Of course the two-fold tax on sexuality must always be a major determinant in optimizing the program of investments. Investment in sexual reproduction must not only be worth the loss of resources from other possible uses, it must be worth the cost of meiosis.

THEORETICAL EXTENSIONS AND EMPIRICAL TESTS

I hope that the reasoning above will be considered plausible, even though it is based solely on an intuitive visualization of the genetics and ecology of a population of persistent localized clones that give off genetically diverse dispersal stages. A number of implicit assumptions have not been justified. I have assumed that environmental gradients behave in such a way that a successful genotype's local fitness has a negligible chance of being higher at any spot likely to be reached by the dispersal stage. Undoubtedly modes of environmental variation could be proposed in which nearly identical conditions, with similar optimum genotypes, could recur rather frequently. The model would not be valid for such an environment. It would not be valid without considerable genotype-environment interaction in determining fitness. This second point is documented to some extent below and in Chapters 7 and 8.

Given these two conditions, the model seems to propose favorable selection of sexuality as a result of negative heritability of local fitness. A high value in one locality is likely to be a peak value, and indicative of lower values elsewhere. I am not clear on whether this feature is essential to the model. Another possible instance of negative fitness heritability is examined in Chapter 5.

The model is certainly related to that proposed in Chapter 2. In its envisioned role of sex in genotype diversification as a way of adapting to new environmental conditions, it is obviously related to models of long-term evolutionary advantages. Rigorous demonstration of essential differences in these proposals, or of the extent to which they are all merely special applications of a general theory, would certainly require much painstaking work by able theorists.

The theoretical work is not likely to be done without preliminary indications as to the kind of theory that is likely to have explanatory value. So despite logical uncertainties, I feel that the model is clear enough to warrant examination of the extent to which theoretical expectation is realized in observation. This chapter will close with some available evidence on the validity of the Strawberry-Coral Model (and to some extent of the Aphid-Rotifer Model), and with suggestions on gathering additional evidence.

Many crop plants show considerable genotype-environment interaction in the determination of measurable characters (discussion and references in Baker, 1969; and Freeman and Perkins, 1971). Environment-variety interaction often accounts for 20% to 30% of the variance. This is not direct evidence of genotype-environment interaction to this extent in determining fitness. A variable character may in fact have little influence on fitness, and if it does, fitness may be a complex function of that character. These uncertainties do not invalidate the assumption that a major source of variability in a character will often be a major source of variability in that character's contribution to fitness. The frequently investigated character "yield" must be closely related to fitness in the environments in which it is measured.

Any field full of vegetatively reproducing plants that also sexually produce seeds at an appreciable rate can provide critical tests of the model. Clones should prove to be localized, with but limited overlap between clonal territories. Well established clones should persist for a long time, but with territories of fluctuating size. Only rarely should a new clone from a sexually produced seedling persist more than briefly.

Undoubtedly the appropriate scales of measurement will vary greatly. Clones of large plants should occupy larger territories and have greater life expectancies than those of small plants. Uniform environments should result in larger territories than heterogeneous ones. Mode and frequency of both asexual and sexual reproduction should influence clonal size, persistence, and distribution. For instance, if there is little or no sexual seed production, clones may be few and widespread. This in fact has been verified (Grant and Grant, 1971; Harberd, 1967). Species that produce an abundance of seeds should consist of more numerous and localized clones. The limited information indicates that this is true in some areas, but that average size of clonal territory can vary greatly over the range of such a species (Harberd and Owen, 1969). Harberd's techniques of clonal identification and analysis of distributions strike me as extremely promising.

34

CHAPTER FOUR

The Elm-Oyster Model

Earlier arguments relate to organisms for which there is no theoretical limit to clonal proliferation of a single genotype. I will now consider organisms that have no clonal proliferation at all. Every physiologically distinct individual is genetically unique and can not be duplicated. It can grow, but there are real if flexible limits on attainable size. There is also a time limitation. Senescence reduces life expectancies for older individuals.

Despite the differences in relation to previous paradigmes, I believe that the essence of the earlier models is still applicable. Elms and oysters have no vegetative reproduction, but adults are enormously variable in size and fertility. More important, a typical adult size covers not one but many distinct habitats that could be occupied by large numbers of competitors during early growth. Competition between hundreds or thousands of seedlings or spat in an area that can support only one adult is much the same as that between adjacent clones of strawberry or coral. While theoretically elms and oysters may die of old age, adult survivorship is greatly variable, and senescence in natural populations is seldom more than a minor modifying influence on age structure (Hamilton, 1966).

This chapter argues that the optimum compromise, between asexual and sexual reproduction in the model organisms, is a zero frequency of the asexual. It assumes that preadaptations for parthenogenetic egg or apomictic seed production are adequate for their rapid acquisition, should they become adaptive. This assumption may often be untrue, especially in animals.

NATURAL SELECTION AND COMPETITION IN
JUVENILE ELMS AND OYSTERS

Selection against incipient asexual reproduction is readily visualized for the elm. Many people live near elms and can see them shedding seeds and later observe the seedlings as they spring up. In a few minutes one can see a single tree stock an area with seeds many times more densely than adult elms could ever exist. The seedlings can locally form almost a lawn of greenery, with a density many orders of magnitude greater than that attainable by adults.

I would regard the space normally occupied by an adult elm as formally analogous to the confined habitats in the Aphid-Rotifer Model. Once seedlings are well established, selection will largely operate, not by the sudden death of one and escape of another, but by one being slightly better able to appropriate resources and minimize losses to parasites and grazers than its neighbor. Individuals are gradually eliminated by prolonged debilitation. By the sapling stage selection becomes almost deterministic, with one of the few best genotypes almost certain to prevail. Chapter 7 reviews observations that justify this view of selection in immature elms.

The genetic outcome of this intense selection would be little altered by having an offspring genotype represented more than once. It would be a wasteful redundancy to stock an adult-size habitat with more than one well-established young of the same genotype. When a tree sheds a hundred or a thousand seeds into an area that can support one adult, the way to maximize the likelihood that the winning seed will be one of its own and not some other tree's is to diversify them genetically.

Three essential elements in the elm's life history make asexual seeds reproductively disadvantageous and account for its exclusively sexual reproduction. The first is its great productivity of seeds, a factor that permits a large amount of selection to take place in a single generation. As noted on page 22, this is a limiting case of the Aphid-Rotifer Model.

The second factor is the proposed genetic selectivity of the competition between adjacent competitors. Slight differences in viability among seedlings and saplings make a great difference in their relative prospects for survival. As an individual wins out over less fit rivals, its increasing size makes it outgrow some problems and grow into others, including competition with more remote individuals that are also the victors of earlier competitions. Before it can reach maturity, an elm must prove itself in a lengthy succession of ecological niches with different sets of conditions and challenges. I would propose that an enormous amount of selection takes place in one generation of elms (Chapters 6–8). The cost of meiosis is merely one of a large number of major selective factors.

The third essential feature is the variable but real ceiling on winnings attainable by even the fittest genotype. A space of 10×10 meters can be fully occupied by a single elm derived from a single seed of the winning genotype. If that genotype had been represented in that space by a hundred seeds it would not have increased the prize at all. Multiple copies of the same genotype would be as wasteful as multiple purchases of the same lottery ticket. If avoidance of this waste gives more than a two-fold advantage, sexual reproduction is always worth the cost of meiosis, and apomictic seed production would be consistently selected against.

Oysters may be small compared to elms, but their larvae are microscopic, and thousands may settle in a space sufficient for only one adult. Among enormous numbers of competitors for a single space, some will be slightly more tolerant of local conditions and more effective at appropriating resources. They will grow more rapidly and crowd out their neighbors. Much mortality results from a seemingly random predation, but the more vigorous oysters rapidly outgrow the predators that take such a large toll in early stages.

There may be a question as to whether larvae from a single female oyster would normally settle with a density of several to many per adult space. A female oyster may produce, conservatively, a hundred million young in an adult lifetime. If 90% of these are lost in the planktonic stage, and 90% of the remainder are lost subsequently to genetically nonselective deaths, the remnant 1% selectively culled would have a density of 1000 per m^2 in 1000 m^2 of habitat, or perhaps ten times tolerable adult density. Bays and estuaries that provide normal oyster habitats may cover millions of square meters, but larvae settle only on favorable surfaces, which would normally be a small proportion of the area of a bay. Also, larvae from a single female would hardly be expected to be distributed uniformly over all attainable attachment sites. I would suggest that diploid parthenogenesis in the oyster could well result in a wastefully redundant stocking of habitat spaces with the same genotype, although hardly to the same extent as in the elm.

SIMULATED COMPETITION BETWEEN ASEXUAL AND SEXUAL ELMS AND OYSTERS

Easily visualized numerical examples are provided by simulations of the following form (Williams and Mitton, 1973). Allow a surface of 100×100 units to be colonized by a hundred propagules. The number of even numbers in a sequence of 100 random digits can be the number of asexual propagules (all the same genotype), and the number of odd entries the number of sexual propagules. Sequences of three digits in a random number table can locate each propagule on vertical and horizontal scales to the nearest tenth of a unit. Each propagule gets a fitness e^{x+y} with x and y representing general and local fitness dosage as in the Strawberry-Coral

Model (pp. 29–30). All asexual propagules would have the same general fitness and get the same value of x. In a large series of simulations, the asexual and sexual propagules would have the same geometric mean of fitness, but the sexual ones twice the variance in total fitness dosage (given that $\bar{x} = \bar{y}$; $\sigma_x = \sigma_y$). After assignment of fitness and position, each propagule attempts to grow 15 units in all directions so as to occupy a circle 30 units in diameter.

With uniform spacing there could be only nine fully developed individuals. With 100 colonists there will be much overlap of prospective occupancy. There are various ways of representing differences in fitness among contenders. Division of areas of overlap can be a function of relative fitness of all prospective occupants, or perhaps only the fittest two or three. When the central point of an individual is overlapped by growth of another of greater fitness, that individual can be considered dead, or any individual with any piece of the surface can be considered viable.

The outcome seems relatively insensitive to the way in which areas of overlap are apportioned, or to viability requirements. The simplest rules, of giving contested areas entirely to the fittest contestant, and of allowing survival with even minute areas of occupancy, is adequate for purposes of illustration (Figure 4). The surface is largely occupied by a few individuals, those that happened to get the highest values of $x + y$. Sexual propagules predominate among these fittest few, because of their greater variance in fitness. I assume that area of occupancy can be equated to genetic success. Thus the sexual propagules have a sizable advantage under the conditions simulated. In the few completed trials the advantage is sometimes two-fold and sometimes less, so that the arbitrarily chosen constants seem somewhere near the threshold of favorable selection of exclusively sexual reproduction.

Natural habitats have two dimensions more often than one, but logically a one-dimensional model is just as valid. A com-

FIGURE 4. Simulated competition between asexually and sexually produced colonists on a habitat surface. Unshaded areas are occupied by asexual colonists, stippled areas by sexuals, crosshatched by neither.

puter program would normally quantify the two-dimensional model by linear scanning, and adjacent transects would be nearly redundant. There would be no advantage over an originally linear model.

For computer simulation Williams and Mitton (1973) randomly scattered propagules on a habitat 1000 units long. Each propagule then attempted to grow 10 units in each direction. Fitnesses were assigned as in the graphic model above, and contested regions awarded to the fittest contestant. Propagule numbers were given by Poisson distributions with the lambdas indicated (Figure 5). It would appear that, under the conditions of competition and fitness variation assumed by the model, the sexual progeny has more than twice the success of the asexual when propagule numbers in the total habitat space are more than about 700, or about 14 per maxi-

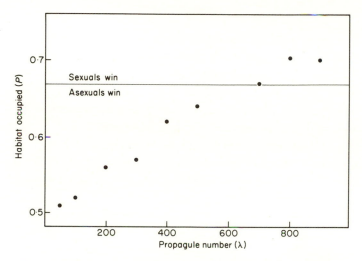

FIGURE 5. Proportion of habitat space won by sexual propagules (P_s) as a function of propagule number (λ) in computer simulation of the Elm-Oyster Model. Each P_s is the mean of 40 trials (from Williams and Mitton, 1973).

mum adult space. In any population for which the model is descriptive and this quantitative requirement is met, sexual reproduction would have a net selective advantage. Parthenogenesis would fail to become established, or if already present would be lost, and an exclusively sexual population would remain.

The model deals with asexual and sexual parents whenever their progenies are in competition. Its validity is unaffected by whether asexual or sexual reproduction is more common.

It applies equally to competition between partly apomictic and partly outcrossed parents. A more than two-fold advantage in the sexual part of the progeny would give a net advantage to the parent with the greater proportion of sexual offspring.

EXCLUSIVELY SEXUAL REPRODUCTION AS AN EVOLUTIONARY EQUILIBRIUM

A sessile animal or plant species might lose the ability to reproduce vegetatively if a genotype could gain more by putting resources into vertical growth from a single stem than from lateral spreading. It would retain exclusively sexual reproduction of widely dispersed propagules if fecundity and variation in fitness were as required by the Elm-Oyster Model. Having lost asexual reproduction entirely, it might also lose or fail to acquire necessary preadaptations for its later reacquisition, even if it became advantageous to do so, a matter explored in Chapter 9.

The degree of within-progeny competition required by the model might be lost, and parthenogenesis favored, by wider dispersal. Smaller seeds may be more widely dispersed. With the assumption that rate of fall through air is inversely proportional to surface:mass ratio, the target area of wind-borne seeds will increase as the inverse square of the linear dimensions of the propagule. Fecundity will increase as the inverse cube of this dimension. So smaller seeds would increase density of colonization and within-progeny competition. Decreased egg size in aquatic organisms in which dispersal depends on swimming powered by nutrients in the egg would result in a smaller target area along with increased fecundity.

The main difference between the Strawberry-Coral and Elm-Oyster Models lies in whether a zygote normally gives rise to a single or multiple "individual." There are plant genera that include some species with and others without vegetative multiplication, and those that have it vary enormously in the extent to which they make use of it. Such genera would be excellent material for testing ideas expressed here, especially the suggested uniformity of ZZI. Monocarpic or single-crown species should normally have greater lifetime seed fecundity than vegetatively produced individuals of closely related species. They should more closely resemble entire clones in their mean output. Their fecundity should provide an estimate of equilibrium ZZI for the group.

Redundant stocking with one genotype may be a population-level analogue of redundant stocking of a community with the same species. Adding large numbers of seeds to a community in which the species is already present need not increase the abundance of that species (Sager and Harper, 1960). There are favorable places for a certain number of adults, and adding seeds merely intensifies competition for the limited number of spaces.

The essence of this and the preceding two chapters is that when many individuals or clones try to complete the life cycle in a space that could be fully utilized by one, selection will allow only the fittest genotype to prevail, and the others will die out. The required high ZZI is achieved by a protist and some simple metazoans with the minimum fecundity of two per generation, but with many generations per life cycle. Oysters and elms achieve their high ZZI in the opposite way, with enormous fertility but only one generation per life cycle. Either way, the high ZZI has important implications that have not yet been considered but bear heavily on the ideas being developed. Their discussion is postponed to the chapters on selection in high-fecundity populations, which advance additional arguments and evidence on the models developed to this point.

CHAPTER FIVE

Other Models

There remain many sexually reproducing organisms that do not conform to any models proposed so far. This chapter will deal with animals that are so mobile throughout life that a permanent close association of sibs, an essential feature of all earlier models, is unlikely. It will be my last attempt to explain sex as an individual adaptation. For all other occurrences, in exclusively sexual low-fecundity organisms for instance, I will appeal to historical constraints that preserve sexual reproduction when it has ceased to be adaptive (Chapter 9).

Sessile organisms and the most freely mobile represent two ends of a continuum. Models developed in relation to sessile forms or those of isolated habitats should be partly applicable to somewhat less sedentary forms, or to those of somewhat larger or imperfectly isolated habitats. In this chapter I will formally regard such animals as starfish and tritons as wide-ranging in extensive habitats, even though many adults may move but little for long periods of time. On piling, one may often find one *Asterias* or *Pisaster* per pile. It may have been there most of its life and be the sole survivor and best genotype for that location. Even for such animals as the cod, factors such as territoriality, homing to habitual spawning grounds, and specialized habitat requirements may make brothers and sisters more competitive than individuals taken at random from the population, and therefore add an element of redundancy to production of genetically identical offspring.

More relevant, perhaps, are limitations on mobility during development. A cod larva or newly settled benthic invertebrate can be regarded as occupying the center of a local neighborhood (limited water mass or benthic area). Density-dependent mortality may operate somewhat independently in different neighborhoods and subject their inhabitants to competitive relations somewhat like those envisioned in earlier models. In outgrowing one such neighborhood, an individual grows into a larger one inhabited by competitors that have proved their fitness in earlier competitions.

Thus there may be some applicability of earlier models to such animals as cod and starfish, but I think that the applicability is limited, and not sufficient to explain the exclusively sexual reproduction of these organisms. Some additional factors must be proposed.

FITNESS VARIABILITY AND SISYPHEAN GENOTYPES

The models to be considered are plausible only with the assumption that selection in populations of high-fecundity animals, even freely motile ones, is more intense, and fitness much more variable, than many people will find intuitively acceptable. It is necessary that some justification of this assumption be provided before the logic of the models is presented.

I will be discussing organisms that, like those already discussed, have a high ZZI and only infinitessimal survival from zygote to maturity. The enormous fecundity permits one to regard most of the larval fish or invertebrates of a water mass as inviable in relation to the sequence of ecological demands that must be met before maturity is reached. The population could persist even if only a small fraction of these larvae had genotypes that would permit survival. This select few would constitute a *genetic elite* in Dobzhansky's (1964A) terminology. It was in relation to studies of low-fecundity Drosophila that Dobzhansky postulated a genetic elite. It is a more credible concept for organisms with a high ZZI.

Genetic elite may be an unfortunate term. It refers to a valid scientific concept that is devoid of moral or social significance. This may not prevent some layman from noting its importance to biologists and using this observation as if it were scientific support for a racially elitist philosophy. Before it becomes better established I would like to suggest substituting the term *sisyphean genotypes*.

Sisyphus repeatedly pushes a boulder up a steep slope until, on the verge of reaching the peak, it goes out of control and rolls to the bottom again. Analogously, an individual in the top end of the fitness distribution has achieved its near maximum of fitness by an only momentarily effective combination of genetics and individual history. The necessarily low heritability of such fitness would probably drop that same genotype down into the range of mediocrity in the next generation. Extremely high fitness would have to be regenerated by some new combination. A definition of sisyphean in my dictionary (*Webster's New International*, 1923) is "requiring continuous redoing."

Sisyphean fitness is a relationship between a limited set of genotypes and a particular succession of environments encountered during development. Dobzhansky proposed that any individual more than two standard deviations above average be recognized as elite. I will use the term sisyphean in the more extreme sense of individuals of many times the average fitness. The existence of such individuals follows from three common assumptions on fitness determination: (1) many different genes contribute to variation in fitness dosage, which can be regarded as continuous and normally distributed; (2) fitness differences per locus are in the range of one to ten percent; (3) dosage contributions are functionally independent and multiplicative. Under these conditions fitness would be a highly skewed lognormal distribution, and an individual's fitness would be proportional to

$$V_i = e^{a+bx_i}$$

where V_i is reproductive value at the zygote stage and x_i is its fitness dosage. The term a would determine absolute fitness and reproductive values and the rate of increase or decrease of the population, while b would measure average departures from multiplicative combination. It would be less than unity and reduce the variance of the distribution, if there were compensatory relations among fitness contributions, thresholds of adequacy, and similar interactions. Crow's (1968) discussion supports the commonness of synergistic interactions that would raise the value of b. The "synthetic lethals" of Kosuda and Moriwaki (1971) would be good examples.

This discussion raises the issues of genetic load, cost of natural selection, and related matters that are currently being warmly debated, and would be impossible to treat adequately in a few paragraphs. Readers unfamiliar with these issues may consult such references as Kimura and Ohta (1971) and Wallace (1970). The position taken here is that a can be as large as is necessary to account for the persistence of the population (the density-dependent form of "soft selection"). The factor b may likewise be less than unity (as in threshold models of soft selection), but I will assume that functional independence of fitness contributions, and consequent high variance in fitness, is a general rule.

The concept of threshold selection (King, 1967; Maynard Smith, 1968b; Sved, 1968) may find some convincing examples in sensory mechanisms. It is plausible that an almost perfectly anastigmatic lens could be produced with a rather low minimum number of favorably selected genes that affect lens symmetry. A somewhat smaller dosage may produce severe astigmatism, with serious reduction in fitness. With this kind of canalization of nearly optimal characters, most of the unfavorable genes that are eliminated would be lost along with other unfavorable genes. Selection of a given intensity would require lower mortality than if fitness were multiplicative.

For other characters, especially for those with optima at infinity or zero, and for relations between functionally different characters, a multiplicative relationship is at least as plausible.

47

In primitive man, visual acuity, malaria resistance, intelligence, fleetness, and fertility must have varied largely independently and affected fitness in a multiplicative way, and each of these characters may be regarded as a complex of several. For instance, resistence to malaria must be determined by several largely independent physiological and behavioral characters. When the data are adequate to distinguish the two possibilities, quantitative characters are often found to have lognormal rather than normal distributions, as would follow from multiplicative interaction (Bliss and Reinker, 1964; Kerfoot, 1969). Bliss and Reinker proposed ecological factors and Jackson (1970) developmental mechanisms that would make lognormal distributions the general rule.

The assumption that complex adaptive characters in nature are widely variable in their contributions to fitness is abundantly justified by performance data. There is no evidence that adaptive performance (mechanical strength, disease resistance, etc.) ever show commonly achieved but never exceeded maxima, as would be implied by threshold selection. Such data in fact may show positive skew. A few individuals may survive a stress for several times the modal duration. High variance and positive skew on one important performance statistic, fertility, will be documented in Chapter 8. A genotypic selection coefficient must be a summary of performance scores, and if individual scores are highly variable, the mean or other general summary should be even more so.

Current ignorance obviously permits a wide range of opinion on the nature and extent of fitness variation in nature, but there can hardly be any support for the idea that there are no genetically and functionally independent contributions to fitness. Even if only a small proportion of the combined effects are multiplicative, they could still generate an approximation to a lognormal distribution with considerable positive skew. Van Valen's (1965) statement that fitness ". . . is known to be negatively skewed with a secondary mode at

lethality and sterility," must relate to the physiologically opti-
mized and predator free environments of laboratory stocks.
In the wild, with natural mortality sufficient to keep the population in check, I assume that most of the individuals in Van
Valen's primary mode would be shifted to the vicinity of the
secondary.

If genotypic fitness in man or Drosophila is determined
partly by multiplicative interactions, the same must be true to
a greater extent in organisms such as starfish, that develop
through a succession of ecological niches. Fitness in bipinnaria
larvae, and in 2-mm, 10-mm, and 50-mm juveniles, and
200-mm adult starfish requires different kinds of adaptations
and their genetic bases should be partly independent. With
complete independence and P_i as the probability of surviving
one stage, the probability of surviving n stages would be
$\Pi_{i=1}^{n} P_i$. This product would be lognormally distributed, even
if the separate P_i's were not.

Figure 6 is an intentionally extreme view of how zygote-to-
adult viability (perhaps the major component of fitness)
might be distributed in a high-fecundity population (at least
millions of zygotes per adult). To regard an individual with
a survival probability of less than one in a hundred thousand
as lethal would be consistent with conventional usage and with
threshold selection models that have been proposed. On that
basis we can regard the whole visible distribution (Figure 6)
as showing only lethal genotypes. Viables, which will account
for nearly all of the next round of reproduction, all lie off
the graph beyond 10^{-5}. Chance largely determines which one
out of hundreds or thousands of these sisyphean genotypes
acutally survives and reproduces.

The validity of upcoming arguments does not depend on
the lognormal being closely descriptive of fitness distribution.
What is required is that fitness have high variance and positive
skew, and that reproductive success depends on having some
offspring of many times the mean fitness. The subsequent

49

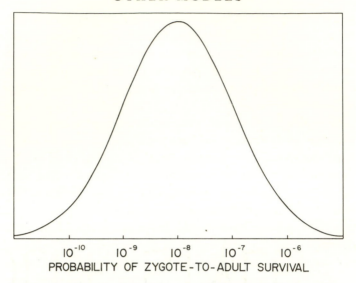

$$10^{-10} \qquad 10^{-9} \qquad 10^{-8} \qquad 10^{-7} \qquad 10^{-6}$$

PROBABILITY OF ZYGOTE-TO-ADULT SURVIVAL

FIGURE 6. Possible distribution of viability in a high-fecundity population. A difference of one standard deviation in genetic fitness dosage causes a tenfold difference in viability. Such viability variation would result from approximately the following conditions: 20,000 loci contribute to variation in fitness; optimal and suboptimal genotypes at each locus differ by 0.05 in viability; probability of suboptimal genotype at each locus is 0.10; adjustment for density dependence of mortality (the term *a* defined on pages 46–47) is 95.

chapters on selection in high-fecundity populations will provide arguments and evidence in support of these assumptions. For the moment I will leave this discussion and proceed with models of selection of sexuality in motile marine animals.

THE TRITON MODEL

It is generally agreed that sexual populations are more likely to survive changed conditions than asexual ones. I suspect that this is true, but for reasons different from those commonly assumed (Chapters 12 and 13). Regardless of the nature of the advantage of sexuality, if its strength is sufficient to produce a major effect in one or a very few generations, it can be regarded as a major factor in selection among individuals of a population. I suggest here that there may be populations in which selection is so strong that a sexual individual, because of the evolutionary potential of its lineage, may have more than twice as many immediate descendants as an asexual. Such populations may be common in shallow tropical and warm temperate seas, and a triton (*Cymatium*) may be a good example.

A triton is a large gastropod that shares the following characteristics with many other mobile benthic invertebrates. Its movements as an adult may be considerable on a local scale, but on a chart of a major part of the ocean an adult habitat can be represented by a point. By contrast, its enormous numbers of young are planktonic for many months, and, ocean currents being what they are, this implies dispersal for hundreds or thousands of kilometers. This dispersal can be strongly directional, because major current systems run in consistent directions. Larvae originating in one locality are carried away and may seldom participate in restocking that locality.

Demographic implications of these long-distance dispersal stages are discussed by Scheltema (1971). Observations on rates of development and on the distribution of developmental stages in the North Atlantic compel the conclusion that dispersal on a major geographic scale regularly takes place. Adult tritons of Florida may have originated as zygotes far to the southeast in the West Indies. Zygotes produced in Florida may reach adulthood much further "downstream," in the Carolinas perhaps. Sexual reproduction in Carolina may contribute to the stocking of Europe and the Mediterranean. The likelihood of transoceanic dispersal is one of the main points of Scheltema's discussion. He reasons that the Gulf Stream is an agent of west-east, and the tradewind driven equatorial current an agent of east-west dispersal.

I envision that colonists at any locality, such as the thoroughly tropical Hispaniola, may be remote descendants of adults that lived on that tropical coast many years before. It was necessary that adaptations to that coast be lost, or strongly compromised, and then evolved anew in the cycle of generations transported by the North Atlantic gyre. If sexual reproduction so facilitates this sort of reversible evolutionary change that a sexual progeny has twice the success of an asexual, the population would become and remain exclusively sexual. The biology and oceanography of the North Atlantic are relatively well known. There are no studies comparable to Scheltema's for other basins, but there is no reason to believe that larval dispersal in the North Atlantic is essentially different from that of other oceans. Similar sorts of current systems and planktonic larvae are found everywhere in warm and temperate seas. A preliminary indication of a parallel to the triton is seen in work on the spiny lobster of western Australia. Larval dispersal patterns make it unlikely that adults of that region originated there as zygotes (Chittleborough and Thomas, 1969).

Some species of oceanic holoplankton must also consist of cyclic chains of geographically separated generations living under different ecological conditions. Most of these are small organisms of limited fecundity, which would rule out the intense selection necessary in the model. Large prolific organisms, in the plankton as elsewhere, tend to be long lived, and may complete such a circuit as the North Atlantic gyre in a single generation. Perhaps a large jellyfish in the slower gyres of the Pacific may have the necessary population structure.

THE COD-STARFISH MODEL

A more generally promising approach, in a quest for favorable selection of sexuality in motile animals, may lie in analysis of breeding structure. There are circumstances under which asexual and sexual reproduction can give different probability distributions of fitness, as envisioned in Figure 1 (page 14), This effect need not follow from greater within-progeny variation from sexual reproduction. This factor was important in earlier models only because of postulated within-progeny competition for what a single individual could fully utilize. Mobile animals of continuous habitats can escape this kind of competition with close relatives. In the present game, ten copies of the winning ticket can get ten first prizes.

With the usual simplifying assumptions of population genetics (Mendelian inheritance, random mating, linkage equilibria, etc.), asexual and sexual reproduction produce the same genotype frequencies and fitness distributions. This is easy to see in a one-locus, two-allele model with a unique phenotype for each genotype. If the parental generation has Hardy-Weinberg genotype distributions, its sexual reproduction will, with the usual assumptions, produce an exactly similar offspring generation. Likewise parental and offspring generations will be exactly the same with asexual reproduction. This population-wide equivalence of asexual and sexual reproduction is true only if the simplifying assumptions are true. If not, the two processes may give different results.

Substitution of assortative mating for random mating is the only modification that I can conceive of being valid for a major part of the world's biota. Application to motile animals must be restricted to highly prolific forms, because only they can sustain the intense selection necessary for the advantage to pay the cost of meiosis. It is equally applicable to prolific sessile forms and could be a factor in addition to those already proposed. The Model assumes a life cycle which, as in the Triton Model, includes a widely dispersed propagule that colonizes a neighborhood. These propagules are dispersed at random, rather than along major current systems. It should be applicable with brief larval stages and in the variable currents of coastal embayments. The neighborhood colonized, in which an individual develops to adulthood, can be large, and is colonized by the young of a large number of parents. Brothers and sisters constitute a negligible fraction of a given individual's competitors.

The extensive movements of an adult cod do not invalidate the neighborhood concept. Migrations are cyclic and residents of a given neighborhood move together and usually return to the same spawning ground year after year (Harden Jones, 1968). A spawning stock may be physically recognizable by fishermen, who speak of "spring races," "summer races," inshore and offshore stocks, etc. Wynne-Edwards (1962) amassed a great fund of evidence in support of the localized breeding-group concept for a variety of animals. I would disagree with his assumption that these neighborhood groups constitute populations in the genetic sense.

Colonists settling in different neighborhoods must already be considerably different. They are a remnant few that have survived the special physical conditions, pathogens, and predator pressures of the water masses that have brought them to different neighborhoods. After establishment in a neighborhood, only a tiny fraction will survive the special challenges of that locality and reach maturity. Phenotypically and genetically the adults of a given neighborhood are uniquely distinctive, not because they are reproductively isolated from other neighborhoods, but because selection is sufficiently intense to generate the distributions anew for each generation. Evidence in support of this interpretation is presented in Chapter 7.

Mating between neighbors therefore means that an individual's mate will usually be more similar to itself than it would be if selected at random from the population. Such assortative mating produces progeny more variable than their parents, because similar parental phenotypes may have been based on a variety of genotypes. This latent variation is then expressed in offspring recombinations. *Assortative mating* may not be an entirely appropriate term, because it normally implies active mate selection, but the result of the phenotypic similarity is the same, whether it arises by active discrimination or the envisioned limits on mate availability. The offspring are more variable than their parents, and more variable than would be produced by asexual reproduction. Pollen vector behavior may result in phenotypic assortative mating in plants (Levin and Kerster, 1973).

A simple and extreme example of assortative mating may help clarify the idea for those not well versed in quantitative genetics. Suppose a population has a character that varies discontinuously from -2 to $+2$, with only integral values being possible, as in charge of an enzyme molecule. Variation is completely determined by variation at two loci of two alleles each, and all gene frequencies are 0.5. Genotype frequencies are in Hardy-Weinberg and linkage equilibria. Relations between genotypes and phenotypes, and relative numbers of genotypes, are as follows:

phenotypic values	zygote genotypes and relative numbers
2	AABB (1)
1	AaBB (2), AABb (2)
0	AAbb (1), aaBB (1), AaBb (4)
-1	Aabb (2), aaBb (2)
-2	aabb (1)

These zygotes are distributed at random among many neighborhoods, each of which permits only one unpredictable phenotype to survive. Table 3 analyzes three possibilities for reproduction among the survivors, asexual reproduction, random mating in a pool of individuals from two neighborhoods, and mating entirely within neighborhoods. All mathematically nonredundant possibilities are shown. Offspring variability from within-neighborhood mating is always at least as great as from either of the other forms of reproduction, and is usually greater. What happens is that asexual reproduction merely repeats whatever variability there is in the parents. Sexual reproduction repeats parental phenotypes, plus others, except for completely homozygous neighborhoods. If the mating can be between neighborhoods, intermediates tend to be produced. This tendency does not occur with within-neighborhood mating.

In nature nothing approaching this extreme example could be expected. Every neighborhood would contain considerable

TABLE 3. Results of three modes of reproduction after neighborhood-specific selection

Parental phenotype demanded		Offspring phenotypic variance		
by one locality	by another locality	from asexual reproduction	from random mating	from mating within localities
2	2	0	0	0
2	1	.25	.38	.5
2	0	1	.84	1.33
2	−1	2.25	1.38	2.50
2	−2	4	2	4
1	1	0	.5	.5
1	0	.25	.71	.84
1	−1	1	1	1.5
1	−2	2.25	1.38	2.5
0	0	0	.67	.67

variability, and only part of it would be genetically determined. Crow and Kimura (1970:155) calculate an increase in standard deviation of about 3% if both heritability and between-mate correlation equal 0.5. The 0.5 may be too high for a typical within-neighborhood correlation. Perhaps 1% would be a more reasonable estimate of average increase in variability per character with populations such as those postulated in the model. Perhaps this is enough, with large numbers of such characters contributing to fitness. Mating within neighborhoods would provide the next generation with greater variability, and therefore greater numbers in the sisyphean range of fitness.

The argument gains plausibility on consideration of properties of highly skewed distributions such as a high-variance lognormal. I am indebted to J. S. Farris for patient instruction in properties of the lognormal distribution. Nothing that I have proposed would alter the genetic fitness dosage or geometric mean of fitness, but even with a constant geometric mean, the arithmetic mean of a lognormal distribution is doubled by increasing the variance by the constant 2 ln2, or about 1.4. It is the arithmetic mean that is scrutinized by selection. This factor was operative in simulations of the Elm-Oyster Model. Assignment of fitness dosage was unbiased, and the geometric mean of fitness was the same for sexual and asexual progenies. Yet the greater variability of the sexuals gave them a greater arithmetic mean of fitness, as indicated by greater occupancy of habitat space.

Figure 7 shows hypothetical lognormal distributions of fitness for asexually and sexually produced progenies. All have the same mean of genotypic fitness dosage, but differences in variance cause differences in mean fitness. Note that with increased variance in fitness (increased intensity of selection) distributions differing by a factor of two come to resemble each other more closely. The greater the variance of fitness, the easier it is for assortative mating to double the arithmetic mean by increasing genetic diversity.

The conclusion follows that selection may suppress parthenogenesis in certain kinds of high-fecundity populations because the variance of asexually produced offspring would be inadequate for generating the sisyphean genotypes on which genetic survival depends. The argument applies to all populations in which mating is so assortative and selection so intense that parthenogenesis would produce a progeny with fitness variance more than 1.4 units below that of a sexual progeny. If fitness is already highly variable, 1.4 may be a small part

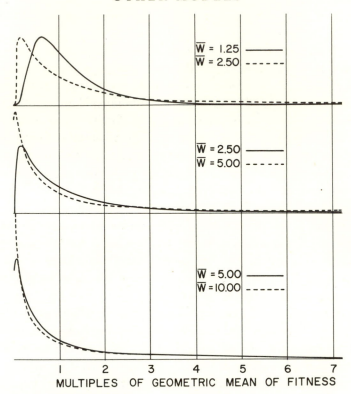

FIGURE 7. Doubling of mean fitness by increased variance of fitness dosage (Cod-Starfish Model). All distributions have the same geometric mean (same arithmetic mean of fitness dosage). Paired distributions differ by a factor of 2 in arithmetic mean and by 2 ln2 in variance.

of the total variance, and mathematically only minor difference between the two modes of reproduction.

The argument has dealt with characteristics of the population as a whole, but does not depend on the success of one population in relation to another. It is not a group-selection argument. All of the frequency distributions for groups are formally valid and have their entire biological significance as probability distributions for individuals.

COMPARISON WITH OTHER MODELS

In the models proposed earlier in this book an asexually produced individual was unlikely to be of sisyphean fitness because, even if it had a winning genotype, there would probably be other individuals of the same genotype with which it would have to share the winnings. No such special within-progeny competition is operative in the Cod-Starfish Model. Here an asexually produced individual is less likely to have a winning genotype. The models are not mutually exclusive. If sperm or pollen transport is much shorter than larval or seed transport, such organisms as oysters and strawberries could conform to the Triton or Cod-Starfish Model.

The models all operate by sexual reproduction generating a variety of new genotypes after selection has produced a generation of specialists for a particular sequence of environmental conditions that has a negligible probability of recurring for the offspring. Thus all show formal similarities to those in which sex is proposed as a long-range group-related adaptation. They merely point out how certain kinds of life histories may give the proposed benefits a hitherto unsuspected importance in a single life cycle. This is especially true of Maynard Smith's (1968A) suggestion that recombination is of value mainly when organisms from different populations of the same species invade a new territory. Colonists may be all AAbb from one population and aaBB from another. It may be that long-term survival is then possible only if most individuals have at least one A and one B. The requisite genotypes would be produced immediately by sexual reproduction. They would probably not have time to be produced at all by asexual. My suggestion is that this sort of advantage, given certain life-history features, can assume great importance in a single life cycle.

Maynard Smith's (1971A) reasoning on sex as an individual adaptation is mentioned briefly in Chapter 1. He found no plausible benefit in sexual reproduction that could even compensate for recombinational load, let alone the cost of meiosis. He found the formal possibility of an advantage when correlations between environmental variables reverse themselves between generations. I take this to be a formalization of the concept of negative fitness heritability. It may thus relate to the Strawberry-Coral or Triton Models. His reasoning is all based on the assumption that to raise fitness one must raise genotypic fitness dosage. He did not consider the possibility of raising fitness by increasing dosage variance.

Of evident relevance to the Cod-Starfish Model is Warburton's (1967) calculation that if fitness varies sufficiently for some genotypes to have more than double mean fitness, selection will increase the variance still further. Higher variance increases the rate at which sisyphean genotypes are produced. Unfortunately, he made the assumption of high heritability of fitness, and infers progress towards fixation of highly fit genotypes and consequent reduction in variance. This exhaustion of fitness variability would not occur if the sisypheans are one group of genotypes in one generation and a largely different group in the next. Selection for variable progenies should continue indefinitely.

There are also the seemingly quite different models of Levins (1968) which show that when selection is strongly disruptive (Levins' *concave fitness sets*) and when parental and offspring environments may differ (Levins' *coarse-grained environment*), selection will favor diversity (Levins' polymorphism or mixed strategy) over any single optimum. My suggestion is that there is commonly a two-fold advantage in such a mixed strategy.

Theoretical work on selection in coarse-grained environments is reviewed by Maynard Smith (1970) and by Cook (1971:110–112). In general they support the belief that diversity and stable polymorphisms are favored by environmental heterogeneity, and that diverse and polymorphic progenies should have a higher level of success than genetically homogeneous ones. These conclusions assume selection acting independently on scattered individuals and are not based on special perils of genetically uniform associations, such as contagious diseases.

Discussions of selection in diverse habitats generally neglect what I regard as the most important life-history pattern for the present discussion, that of widely dispersed young that attempt to colonize sedentary adult neighborhoods. Cook (1971) maintains that his discussion is relevant to sessile marine invertebrates, but clearly it is not, because he assumes random mating among adults. In extreme form the Cod-Starfish Model would have young from each habitat niche randomly dispersed to all such niches. The population would be perfectly panmictic in that each zygote has an equal probability of mating with any other zygote, but once established in a neighborhood, and subsequent to selection, it can mate only within the neighborhood. Levins does discuss this sort of life history, but only briefly. It clearly warrants detailed exploration.

CHAPTER SIX

Natural Selection in
High-fecundity Populations: Theory

This chapter offers a priori reasons for believing that the intensity of selection in a population has a positive relation to its fecundity. Chapters 7 and 8 offer empirical evidence for the same conclusion. Both arguments and evidence relate mainly to organisms that have a high ZZI as a result of adults being highly prolific. So the main relevance is to Chapters 4 and 5. The proposition that in these organisms a tremendous amount of genetic change can occur in a single generation requires special substantiation. I assume that no special support is needed for the idea that considerable genetic change can occur in a single life cycle that includes a long period of competition among clones.

For purposes of discussion I will divide the spectrum of fecundity into three ranges. Low-fecundity organisms are those with fewer than 10^3 zygotes per average adult female lifetime. Those with between 10^3 and 10^6 lifetime output are of medium fecundity. High-fecundity organisms have more than 10^6 zygotes per adult female. *Fertility* will refer to individual variation in production of young at the stage of initial independence from parents (Chapter 8).

It is fair to say that current information and thinking on ecological genetics are biased towards low-fecundity organisms: mammals, birds, fruit flies and other insects, crop plants and weedy annuals. Until recently there was little information on the population genetics of high-fecundity organisms: large trees with small seeds, large fishes and invertebrates with small eggs, macroscopic lower plants with microscopic spores. I believe that it is only in these organisms, and in those in which clonal proliferation of genotypes can take place, that sexual reproduction can be adaptive and currently in evolutionary equilibrium with asexual. I believe that if evolutionary theorists had not been so preoccupied with low-fecundity organisms with simple life cycles, they would have long ago seen the significance of sexuality.

FECUNDITY, THE STRUGGLE FOR EXISTENCE, AND THE INTENSITY OF SELECTION

Selection in relation to fecundity can be approached instructively by forgetting for a moment the abstract genetic meaning of fitness and considering it as an immanent physiological property of an organism in relation to immediate problems. If we take a large sample of a population and subject it to a severe stress, we may reach a point at which half the sample has succumbed. Almost certainly there will be some genetic difference between the dead and the living. This is the sort of thing most biologists who are not theoretical geneticists think of as differential fitness.

If the stress did have this genetic bias, it can be described as a selective event. Ultimately the rate of selection is determined not by absolute time or by generations or by individual deaths but by the flux of selective events. Where there are many selective events per generation, selection will have marked effects in one generation. As an example, suppose the (perhaps unlikely) proposition that juvenile deaths in Pleistocene man and cod were equally selective genetically. In man perhaps half of normal births survived to puberty. The mortality can be attributed to a sequence of stresses, each of which removes 10% of an age cohort. It would take 7 such stresses to produce the requisite killing. Now suppose there is a genotype that has such increased resistance to each stress that only 9% are killed. With 7 such stresses, there remain at puberty only $0.9^7 = 0.48$ of the original group, but $0.91^7 = 0.52$ of the favored genotype. With both types equally represented at birth, the fitter would have become about 8% more abundant by puberty.

Now suppose that there were two such differently selected types in a cod population with, conservatively, 10^7 zygotes per adult female per lifetime. The required 153 episodes of stress would give $0.9^{153} = 1 \times 10^{-7}$ survival of the more vulnerable, and $0.91^{153} = 18 \times 10^{-7}$ of the less vulnerable. With equal selectivity of deaths, the rate of selection is a simple function of fecundity, and selection could accomplish in one cod generation what would take perhaps 25 human generations. Equal selectivity of deaths would mean that the appropriate unit of evolutionary time would be cohort half-lives or some related unit.

The classical Darwinian argument proceeds by first establishing that, because of the tendency of each species to increase beyond environmental limits, there is a "struggle for existence." With different individuals being differently endowed by heredity for this struggle, there must be a natural selection whereby the fit increase and the less fit decline in abundance. Many discussions imply that intensity of selection is somehow related to the intensity of the struggle for existence (e.g. Gause, 1934; Nur, 1970; Emlen, 1973). The struggle is surely greater in the cod, where one in ten million can survive, than in man, where the odds are about even. Yet on the rare occasions when the possibility of carrying this argument through to its obvious conclusion is considered, it is rejected on the basis of unsupported statements that the high mortalities of prolific organisms are nonselective genetically.

Recent examples of this rejection are provided by Mayr (1962), and Wynne-Edwards (1962). An earlier example is Watson's (1936), whose introduction to a symposium proposes that

. . . a heavy non-selective infantile death-rate, such as that which occurs generally in marine teleosts, merely reduces the population within which selection of adults may operate.

No evidence is offered in support of the nonselectiveness of infantile deaths in any organism. Perhaps the thinking behind such statements is based on the idea that natural selection is mainly concerned with making things evolve. Man has evolved more than most other organisms. Since man's lineage has included nothing but low-fecundity organisms for perhaps a quarter of a billion years, it must follow that low fecundity does not reduce the strength of selection.

It would be more correct to say that natural selection is mainly concerned with preventing evolution, not causing it. Much selection is concerned with the elimination of low-fitness genotypes produced by mutation or recombination. It can be at a generally intense level, but vary so in direction and strength at different times or places that little cumulative change takes place (Dobzhansky, 1964b). Even in a rapidly evolving population, only a minor part of the force of selection need be concerned with more or less permanent gene substitutions. Man's lineage may have been evolving more rapidly than the cod's all through the Cenozoic, but this need not rule out the possibility that the cod lineage experienced more intense selection.

Watson's statement introduced its own rebuttal. A contribution to the same symposium (Salsbury, 1936) reviewed competition and selection among seedlings and concluded that their heavy mortalities are highly selective genetically. This conclusion is reaffirmed by Harper (1965a and b) and Stebbins (1970).

HOW SELECTIVE ARE HIGH JUVENILE DEATH RATES?

While I think that the genetic selectivity of early attrition in high-fecundity organisms has been generally underesti-

65

mated, I can scarcely urge the opposite view that all levels of mortality are equally selective. In a cod-man comparison, it is easy to imagine that a young cod's planktonic world is more governed by random events than the world of cave and campfire and well-trodden trail. A cod embryo drifts, slightly buoyant, in the upper strata of the open sea from one to four weeks, according to temperature. During this passive drift it may or may not find its way into the mouth of a predator. When newly hatched the sensory and motor mechanisms are still rudimentary, and the larva shows only two kinds of behavior, seemingly random swimming at a constant rate, and not swimming. Environmental discrimination of the fit from the less fit may seem unlikely.

But there may be reason to doubt the reliability of intuition in judging selectivity of death among cod eggs and larvae. If egg buoyancy is genetically variable, different genotypes will develop at different average depths and be subject to different conditions, which could have considerable annual variation. Locomotor and sensory mechanisms of early larvae may be of little use for flight from a specific danger but should suffice for depth selection in relation to light and other variables. Genetic differences in habitat selection could certainly result in different predator susceptibilities. It could influence success in feeding, and this has been recognized as an important direct or indirect factor in larval survival (Beverton, 1962; Blaxter, 1965). Not all death is from predation. Might not genotype influence susceptibility to bacterial infection or physical stress? Before the use of antibiotics, bacterial attack largely frustrated laboratory studies of marine fish development. Lastly, any hazard that can be outgrown or grown into will be selective in relation to rate of development.

The conclusion seems justified that not much is understood about the relationship of fecundity to the intensity of selection. If deaths in man and cod are equally selective, and there is close correlation between fitnesses of successive developmental stages, the amount of selection per generation would be proportional to the number of cohort half-lives per generation. A gene frequency could be changed as much in one cod generation as in 25 human. With no between-stage correlation in fitness, the amount of selection would vary as the square root of the number of cohort half-lives, or about five times as much per generation in the cod. If it is felt that mortality must be somewhat less selective in the cod and that there would be some between-stage correlation in fitness, perhaps these factors would cancel and five would be a reasonable guess. Evidence reviewed in Chapter 7 suggests that this would be an underestimate.

No matter what an individual's age, when it dies the event is recorded as a change in gene frequency at every locus at which there is variation. When another dies, the original information is amplified to the extent that there are special resemblances between the two individuals, but partly erased or reversed at those loci at which they differed. The fact that whole genotypes die, and not merely genes, makes natural selection a wasteful way of generating information. It takes many deaths to produce an appreciable amount of selective change. Just how many is currently a matter of controversy, as indicated in the last chapter. When the debate relates to real organisms, it usually refers to low-fecundity forms like man or Drosophila. Note that King's (1967) graphic model of low-cost evolution has selection weeding out the lower end of a fitness distribution and allows the majority equal odds in the lottery of survival. My suggestion for high-fecundity organisms puts the threshold in the opposite tail of the distribution.

The important point for the present discussion is that high fecundity, in providing larger numbers of genetically different individuals, allows for more genetically selective deaths. Man's "fertility excess," according to Nei (1971) permits about one gene substitution in ten generations. The fertility excess of cod or elm must be sufficient to permit the equivalent of many such changes in a single generation.

FITNESS FOR THE ENVIRONMENT AND FITNESS OF THE ENVIRONMENT

The supposition that ordinary genotypic variation can have an enormous influence on survival gains credibility from great differences in survival caused by ordinary environmental variation. In high-fecundity marine fishes the spawning of different years produces, independently of initial egg numbers, widely variable numbers of adult survivors. Ten- to hundredfold variation among year classes is common and a 15,000-fold difference is documented (Sette, 1961). This is merely a carefully studied example of the common impression that a year class can sometimes be missing entirely. It might be said that all the zygotes produced in some years have lethal genotypes in relation to the environments encountered. If ordinary environmental variation can produce such widely different rates of survival, is it not reasonable to suppose that ordinary genotypic variation can do the same? An alternative statement of the same idea is that fitness is not a character that can be canalized in relation to adverse genetic factors, any more than against adverse environmental factors.

Great variation in year-class strength is especially characteristic of high latitudes where annual environmental variation is great. The conditions responsible for good and bad year classes interact in complex fashion, and fishery biologists have not been very successful in predicting year-class strength from simple measurements such as temperature or wind direction during development. Great variation in year-class strength of prolific freshwater fishes (pike and larger percids) is also common and attributable to environmental variation early in development (Elrod, 1969; Noble, 1972; Walberg, 1972).

Large fluctuations are especially characteristic of species

68

(both fishes and invertebrates) with the greatest fecundity and longest planktonic stages (Coe, 1953; Thorson, 1950; A. B. Williams, 1969). Great year-class variation for several species of forest trees was noted by Hett and Loucks (1968) and the variation by the end of the first year was largely independent of initial seed abundance. Salisbury (1942) maintains that the more prolific herbaceous plants show greatest variation in abundance. Annual poppies are a striking example (Harper, 1966).

Figure 8 shows my conception of how genetic and environmental factors produce irregular variations in survival. I assume that survival is often by chance, so that every individual has only one chance in a hundred of avoiding a random, non-selective death. I also assume genetic deaths of widely variable probability. This variation is caused by differences in genotypic fitness dosage that are normally distributed and multiplicative. Each standard deviation in dosage produces a manifold difference in viability.

The extreme years ($n + 1$ and $n + 2$) are only two standard deviations apart in fitness dosage, but this means that a hundred times as many will survive in $n + 1$ as in $n + 2$. Survival depends on how far the extreme right end of the distribution reaches into some high (sisyphean) range of survival probability, such as 10^{-5}. This view is in conflict with some recent work on fitness in relation to genetic and environmental variation. Threshold selection models such as that of King (1967) imply that only those below a certain dosage threshold will have markedly variable prospects for survival. This model represents fitness as a character developmentally buffered against both genetic and environmental variation.

That widely fluctuating populations seem not to have a high rate of extinction implies that a zero rate of survival is approachable but is seldom actually reached. We would expect a positive skew in the frequency distribution of year-class strength. If variation in positioning of the curves (Figure 8)

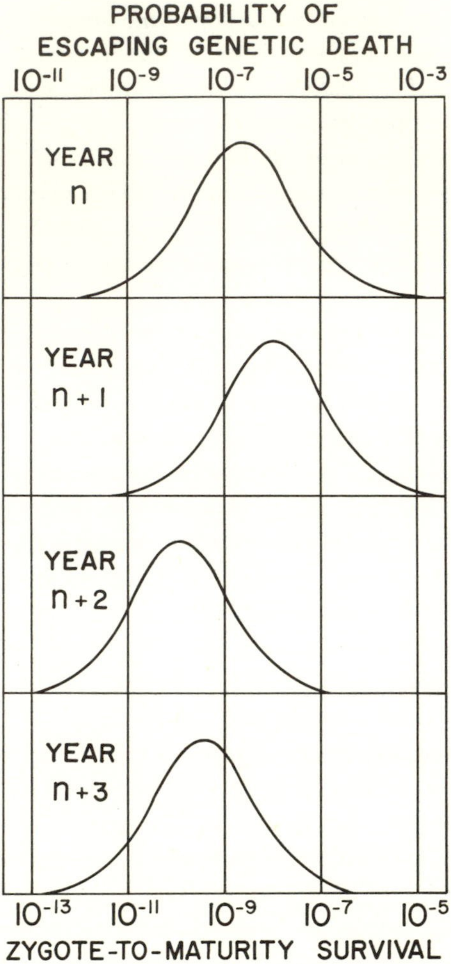

FIGURE 8. Interaction of genetic and environmental variation in producing differences in year-class strength in a high-fecundity population. Values on the upper scale are a hundred times greater than those on the lower, to reflect a high frequency of random mortality. Genetic variation is shown by the spread of the curve and is assumed to be similar each year. Variations in habitat suitability are indicated by shifts of the entire curve along the survival axis.

is normally distributed, survivorship would have a lognormal distribution. Examination of published data, which unfortunately seldom provide large samples of year-class values, gives some support to the proposal that these values have lognormal distributions (Figure 9). Southwood (1967) considered theo-

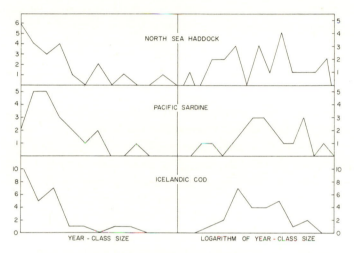

FIGURE 9. Frequency distributions of year-class strength in marine fishes. The normalizing effect of logarithmic transformation supports the inferences of (1) functionally independent (multiplicative) effects of environmental factors in determining survival, (2) a lognormal distribution of environmental suitability, and (3) lack of canalization of the character fitness.

Top curves are from measurements on a graph published by Graham (1956). Middle curves are from data tabulated by Clark and Marr (1956). Bottom curves are from measurements of catches of 8- to 11-year-olds graphed by Jónsson (1957). Two of the data points are omitted from the logarithmic plot for the sardine. They are for year classes following the "collapse" of the population and its replacement by an anchovy. These year classes would lie far to the left of the distribution shown. The transformation emphasizes the extraordinary nature of this collapse.

retical aspects of regulation in widely fluctuating populations, and found that permissible fluctuations are related to fecundity, although great numerical variation is possible with ZZI as low as 50.

A lognormal distribution of "fitness of the environment" follows logically from a lognormal distribution of "fitness for the environment" on the assumption that environmental fluctuations have proportionately equal effects in different parts of the fitness distribution. Arguments for functionally independent (multiplicative) fitness contributions and the resultant lognormal distribution of fitness are much the same whether the contributions derive from genotype or environment.

DEMOGRAPHIC IMPLICATIONS OF HIGH FECUNDITY

High-fecundity populations can be thought of as non-Markovian, as opposed to Markovian populations of low fecundity. The human and other low-fecundity populations are Markovian in that the size of a given generation is largely determined by the size of the preceding one. Evidence reviewed by Southwood (1967) indicates that this is generally true of low-fecundity insects. The size of the next generation of cod or elm trees is not so determined. It will be almost independent of present numbers or anything else in recent history and almost entirely determined by the current size and suitability of habitat.

This distinction has a number of important consequences, including some practical ones. Life tables and related Markovian models are of no use for non-Markovian populations. In low-fecundity organisms destruction of young or interference with reproduction may provide control of numbers. Release of sterile males proved an effective means of control of some low-fecundity insects. It would be ineffective against some of the more important weeds, undesirable fishes, etc. Likewise the artificial introduction of young may be an effective method of augmenting a low-fecundity but not a high-fecundity population. The widespread establishment of marine fish hatcheries in the last century was based on a Markovian concept of fish populations. Their almost total abandonment in the present century reflects the ascendancy of non-Markovian models (Beverton and Holt, 1957; Ricker, 1954; Gulland, 1971).

It should be easily acceptable that a population with extensive clonal multiplication would be non-Markovian on a time scale of entire life cycles. Suppose, for example, that you were told that two protozoan cultures were started two months ago, one with 10 individuals, the other with 100, but the records are lost as to which was which. Now one of the cultures contains a thousand individuals and the other a million. Clearly this information is useless in the problem of deciding which culture was started with 10 and which with 100. Similarly, if you were told that the 1970 spawning of a cod population produced 10^4 mature individuals and the 1971 spawning 10^6, there would be no basis for saying anything about the sizes of the original innocula of zygotes. Perhaps there were about 10^{13} zygotes both years, each one a potential adult. Perhaps also a single protozoan could produce 10^{13} in two months. In both cases, the accumulation of history obliterated all traces of original potential, but this was accomplished in perhaps 40 generations in the protozoan, and only one in the cod.

The cod is one of the best studied examples of a high degree of independence between spawning stock size and numbers surviving from a given spawning. It is capable of living many years and growing to an enormous size and fecundity, many times greater than those of a newly matured individual of a few years. Today there are productive cod fisheries that depend largely on juveniles, because hardly any survive to sexual maturity. Spawning adults represent only a small fraction of normal adult numbers, and they consist largely of first spawners, with only a small fraction of attainable adult fecundity. Reduction of the spawning stock by fishing must curtail the production of cod zygotes as drastically as any release of sterile males has ever curtailed production of fruit fly zygotes. Yet the remnant cod population is capable, every few years, of producing enormous year classes.

Older juveniles of yellowfin tuna are also large enough to enter the fishery, and even average-size year classes may be almost completely removed before they have time to mature (Davidoff, 1965). Yet there is no suggestion that curtailment of zygote production has any adverse effect on numbers of yellowfin tuna.

Cushing (1971) recently produced a thought-provoking analysis of the stock-recruitment problem. Manifold variation in recruitment from a given stock size is universal in all species, but Cushing found that by combining data from many studies of a given species he could get average relationships of some significance. Recruitment is positively correlated with parent stock size in species that produce mere thousands or tens of thousands of eggs per female per year. For species that commonly lay eggs in the hundreds of thousands, stock and recruitment seem completely independent. For species with fecundity in the millions, such as cod and tuna, a negative correlation becomes apparent. Evidently there is some over-compensating density-dependent effect in these high-fecundity populations. In my view the disappearance of the positive correlation between fecundities of 10^4 and 10^5 (roughly 10^5 to 10^6 per female lifetime) documents the transition from slightly Markovian to non-Markovian populations.

It would be of great value to carry out such studies as Cushing's with other fishes and other groups of organisms. The alewife is only a medium-fecundity fish, but Brown (1972) found slight negative correlations between its stock size and recruitment over a period of several years. Harper (1966) found that seed production and recruitment in medium-fecundity poppies were largely independent, but that there was some tendency for intermediate levels of seed production to produce greater numbers of recruits.

An old adage among microbiologists holds that all possible

microbes are latent everywhere, and all one need do is supply a certain environment and appropriate microbial species will appear in appropriate numbers. Van der Pijl (1969) suggests that this idea may be applicable, with obvious limitations, to prolific higher plants. I would also suggest applicability, not only to the species composition of a community, but also to genotypic composition of a species. A given environment will select its sisyphean few from the vast pool of available genotypes. Gene frequencies of these select few may prove quite different from those in the original pool. Diversification of genotype among widespread propagules produces augmented numbers of sisypheans and is especially adaptive in non-Markovian populations.

THE SEARCH FOR EVIDENCE

The proposition that high-fecundity organisms commonly experience great genetic change in a single generation is obviously, if not always easily, testable by data from field and laboratory. Measures of adaptive performance in elm seedlings, oyster larvae, etc., should show considerable individual variation. The evidence should support the assumption that different genotypes have markedly different likelihoods of surviving a given episode of stress or of exploiting opportunities. Physiologically measured inbreeding depression should be considerable. Response-to-selection experiments should generate great changes in a single generation. Not only should selection for a given character be able to accomplish significant progress in one generation, but indirect selection regimes should also cause significant change. For instance, the progeny of a pair of oysters raised in environments that differ in temperature, salinity, food organisms, kind of predators, should show markedly different gene and genotype frequencies at some loci after the high mortality stages are passed.

Individuals from wild populations should be found to differ

greatly in measurable kinds of adaptive performance. For in-stance, there should be great fertility variation. Field studies of high-fecundity species should show marked shifts in gene frequency between localities, even when gene flow is so great as to approach panmixia. Environmental fluctuation in time should be closely tracked by genetic changes among inhabi-tants of the changing environment. This should be especially observable among annual species at high latitudes. Different year classes at the same locality should show different gene and genotype frequencies. The amplitude of genetic fluctua-tion should parallel that of environmental fluctuation. Within a year class at a single locality, marked departures from Hardy-Weinberg equilibria should characterize some loci. The next two chapters summarize evidence on some of these points.

CHAPTER SEVEN

Natural Selection in High-Fecundity Populations: Evidence on Viability

This chapter presents two kinds of evidence on the generality of large fitness differences in high-fecundity populations. The first documents individual variation in physiological measures of viability in stages with heavy mortality. Survival must be influenced by the genotypic component of this variability, and this influence could be felt many times during the development of a cohort through high-mortality stages.

The more important kind of evidence relates to genetic gradients in relation to dispersal. If nearby localities show marked differences in gene frequency in a widely dispersed species, it must mean that powerful selection is operating. The same conclusion follows from genetic variation among age cohorts at a single locality.

VARIATION IN ADAPTIVE PERFORMANCE

There is a large mass of poorly documented scientific folklore that bears on variation in physiological measures of viability in high-mortality stages. Those who grow seedlings or fish fry commonly find great differences in size and other superficial signs of vigor. If any considerable fraction of such variation has a genetic basis, great genotypic variability in fitness must be frequent. Indications of the genetic origin of much of the variation in vigor are discussed in Chapter 8.

Recently the progress of size variation and mortality has been observed in detail for experimental plantings, and confirming evidence gathered for wild populations. The data relate to a variety of plants, from small herbs to forest trees. The account below is a summary of the work of Koyama and Kira (1956), Bliss and Reinker (1964), Stern (1965), Risser (1970), Hett (1971), Hett and Loucks (1971), and especially White and Harper (1970) and Ross and Harper (1972).

Seeds and seedlings commonly show an approximately normal size distribution and modest variability. In nature, and in crowded experimental plantings, variability increases with age, and the distribution grows more askew. The asymmetry gets so great that Koyama and Kira speak of an L-shaped size distribution. The high variability and skewness result from amplification of what are originally slight differences in individual vigor and in the local level of crowding. An individual that has a slight initial advantage, in the occupation of what Ross and Harper call "biological space," can use that advantage to win additional advantage.

During growth of an age cohort there is steady mortality, with the rate per unit time decreasing as time passes. Consistently it is the smaller individuals that suffer heaviest loss. This conclusion is supported by direct observation, and White and Harper calculated that a steady culling of the smaller individuals must be assumed to account for the observed changes in size and number through time. The competitive relations during development and the size bias in mortality are precisely those postulated for the Elm-Oyster Model. The final survivors must be those at the top of the distributions of both general individual fitness and of habitat fitness for the individual. The size differences also have implications for fertility variation, which is discussed in the next chapter.

I am sure that the intuitions of oystermen and fish culturists would support the generalization that similar developments in size and mortality are characteristic of crowded groups of juvenile animals, but I know of only one careful study. Rose (1959) showed that size variation rapidly increases during tadpole development and that smaller individuals suffer greater mortality. Chemical contamination of the habitat by the larger individuals was the responsible factor. Population densities in natural tadpole habitats are great enough for the effect to be important in nature. Comparable phenomena among fishes of open waters must have some other immediate cause.

Besides the steady attrition often seen in a developing cohort, there are sometimes mass mortalities, such as "damping off" in a tray of seedlings, or comparable calamities among fish fry. The mortalities may result from fungal or bacterial infestations or physical stresses that kill the great majority but leave others apparently unaffected. If even a small fraction of such deaths are influenced by genotype, such an event could greatly change gene frequencies in an age cohort. Yet there could still be thousands of individuals left from a single pair, and this group could be subject to many more selective events before maturity.

GENETIC LOAD IN HIGH-FECUNDITY POPULATIONS

Absence of a relationship between fecundity and selection, as implied by those who claim that heavy juvenile mortalities are nonselective, implies that selectivity of death at a given early developmental stage must be quite low in a high-fecundity population. It would be expected that demonstrably low-viability genotypes, or any indication of inbreeding depression, would be harder to demonstrate at an early stage of development in a high-fecundity species. On the other hand, if deaths are as selective in high-mortality stages as in others, they should show no great uniformity in measures of viability.

Crumpacker (1967) reviewed evidence of genetic load and inbreeding depression in a great variety of plants and animals of widely variable fecundity. There is no indication that low-viability genotypes are infrequent in the higher ranges of fecundity. Most of the evidence relates to cultivated forms or experimental plantings, but any strong relation between viability variation and fecundity in nature should be demonstrable to some extent in artificially maintained wild stocks.

Marked variation in viability has been demonstrated for embryos of natural populations of loblolly pine (Franklin, 1972) and for seedlings of feral Scotts pine (Bannister, 1965). The seedling variation was caused by a lethal gene for "fused cotyledons" that was common enough to cause considerable inbreeding depression and suggest heterozygote advantage as the reason for its persistence. No simple genetic basis for spontaneous embryonic mortality was apparent for the loblolly pine. Sorenson (1969) compared selfed and outcrossed Douglas fir embryos and found a high frequency of embryonic lethals and a level of inbreeding depression at least as high as in man or Drosophila. It would be unreasonable to assume, where lethal genotypes are common, that those of partly reduced viability would be uncommon. It is also unlikely that great viability differences would exist in the embryonic but not in later stages, but these points remain to be established.

Inbreeding depression in the fern *Pteridium aquilinum* can be so great that the species was considered to be self-sterile. Klekowski (1972) showed that this is not so, but that some specimens have so many recessive lethals that selfed gametophytes fail to yield any viable sporophytes. Other individuals do produce limited numbers of sporophytes through selfing. The genetic load varies widely among localities, but is generally much higher than in man or Drosophila.

The limited evidence on early-acting lethals and inbreeding depression in high-fecundity populations certainly supports the assumption of considerable viability variation during high-mortality stages. Valuable data could be obtained from a variety of high-fecundity animals. For instance, it would now be routinely feasible to measure genetic load in oysters and plaice by comparing brother-sister with outcrossed matings. The occasional protandrous hermaphrodites and possibility of storage of frozen sperm make selfing possible in the oyster (Lannon, 1971).

LOCAL DIFFERENTIATION IN MARINE FISH POPULATIONS

The quantity of interest is variance in genotypic selection coefficients in nature, undoubtedly one of the more elusive of the field biologists' list of elusive quantities. The measurements are needed for little-studied high-fecundity populations. It should not be surprising that the evidence has to be indirect. My strategy is simply to establish that gene frequencies differ among nearby localities. *Nearby* must be in relation to dispersal powers and resultant gene flow. Differences and distances must be such that large one-locus selection coefficients must be postulated to maintain the gradient.

Before considering the fishes, it is worth noting that there are indications of marked one-generation changes in extreme examples of low fecundity. Artificial changes in gene frequency, caused by introduction of genetically different individuals, were rapidly restored to normal in a squirrel population (Voipio, 1969). Fine-grain spatial variation in house-mouse populations was attributed by Selander, Hunt, and Yang (1969) to selection. Marked cyclical changes in gene frequency occur in voles along with cycles of population density (Tamarin and Krebs, 1969). Clarke, Dickson, and Sheppard (1963) estimated fitness differences of 50% or more for color variants of a butterfly. The most viable form in one environment could be nearly lethal in another. All these studies relate to one or a few genetic loci. Whole-genotype variance in fitness must be large, even in these low-fecundity populations.

The most abundant genetic data on high-fecundity populations are on marine fishes. The studies make use of easily identified components of blood serum, liver cells, or other fresh material. Electrophoretic or other techniques demonstrate that

81

one or another or both of a pair of similar substances may be present in a single individual, in parallel with expected homo- and heterozygotes. Phenotypic proportions in a sample usually approximate a Hardy-Weinberg distribution. Similar enzyme variants can be shown to behave as Mendelian unit characters in Drosophila and other organisms. It is reasonable to assume that each substance is the product of a particular allele at a unique locus. Given adequate samples, it is a simple matter to estimate a gene frequency of, say, cod at a given station on the Norwegian coast, and to compare it with similar estimates from farther north or south, or from other times at the same locality.

A considerable mass of information is reviewed by Ligny (1969). The best studied species is the European cod, for which Møller's (1969) map of a hemoglobin frequency off northern Norway is shown as Figure 10. Cod eggs and larvae

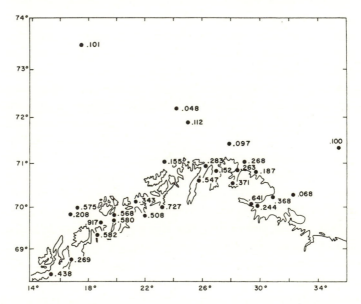

FIGURE 10. Frequencies of cod blood type E off northern Norway (from Møller, 1969).

are transported by ocean currents from one to several months, depending on temperature, and transport for hundreds of kilometers is common (Marty, 1965). Even if the later juvenile and adult stages were completely sedentary, we might well expect gene frequencies to change but little over hundreds, perhaps thousands of kilometers. In fact they can be strikingly different in much less than 100 kilometers. The steepness of the gradients can only be explained by postulating large (10% or more) one-locus differences in selection coefficients between localities. Manifold whole-genotype differences must be common. Many similar examples are reviewed by Ligny, and occasionally there is evidence as to what environmental factor is responsible for the selection. Johnson (1971), for example, showed a temperature dependence for an enzyme polymorphism of a blenny.

It is usual for fishery biologists to attribute gene-frequency differences to spatial isolation of stocks, and it is stock definition that motivates the studies. That the differences are maintained by adults of different localities shows that neighborhood groups do remain separate, as required by the Cod-Starfish Model. This is all a fishery biologist needs to know for stock definition. A population geneticist needs to know more. Genetic differences between localities reflect the joint action of selection and isolation. Any deficiency of isolation, at any stage of development, can be compensated by greater selection pressures in maintaining a gradient. With adequate selection, any difference in gene frequency could be maintained, even if adults were only partly isolated, and the young not at all.

Lack of appreciation of the power of selection in a high-fecundity population led Møller to what I regard as an untenable conclusion. He noted that near-shore and off-shore samples differed consistently in gene frequency, despite the often short distances involved. These distances seemed so minor in relation to the dispersal potential that he reasoned that only an intrinsic isolating mechanism could account for the difference. So he proposed that near-shore and off-shore populations belong to different sibling species. I would explain the differences in the opposite manner, and propose that there is no genetic isolation at all. Both near-shore and off-shore areas derive their adult cod from the same pool of planktonic young. Genetic differences result from the shallow-water areas selecting one limited group of genotypes from the pool, and the deep-water areas selecting another.

The ideal fish for resolving this controversy is the catadromous eel, either European or American. Ocean currents bring planktonic young from spawning grounds in the tropical Atlantic to estuaries from the Arctic to North Africa and the Caribbean. There has been some controversy as to whether the European and American forms are really different, but there is no important challenge to the assumption that in either of these areas there is a single more or less panmictic population, all derived from the same Sargasso Sea spawning. Any genetic differences between localities must be attributed to effects of selection in a single generation.

Marked variation has been found in electrophoretic patterns of various proteins in the European eel from a variety of localities from Iceland to Greece (Drilhon, et al., 1966, 1967; Panteluris, et al., 1971). Unfortunately the published data are not consistent with a genetic interpretation (Koehn, 1972), and can not be used to derive gene frequencies. They are at least suggestive of genetic differences between localities.

More convincingly genetic data are available for the American eel (Williams, Koehn, and Mitton, 1973). Samples from five widely spaced localities from Florida to Newfoundland, a latitudinal range of less than half that known for the species, showed marked differences in gene frequency between localities. At three of the gene loci studied there was a clear latitudinal trend (Figure 11). Other loci varied significantly between sampling sites, but not in any simple pattern. Large, single-locus fitness differences must be invoked to explain the

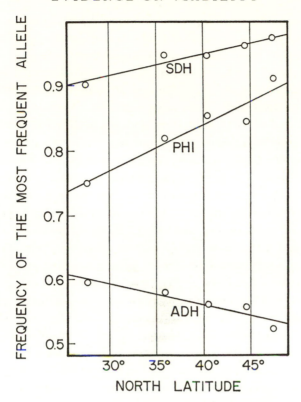

FIGURE 11. Observations and least-squares regressions of gene frequency as a function of latitude for sorbitol dehydrogenase, phosphohexose isomerase, and alcohol dehydrogenase. Regressions are significant at the 0.05 level or lower. From Williams, Koehn, and Mitton (1973).

production of such geographic variation in a single generation of selection. The loci examined must be typical of hundreds or thousands in the population. There must be an enormous variation in fitness among genotypes in the eel.

Alternatively, geographic variation in this species may suggest a need to reexamine the evidence on which the single-population theory is based. Wynne-Edwards (1962) raised doubts, and suggested that despite the oceanic spawning and wide dispersal, there may be some factors operating to preserve the genetic identity of local stocks. Unfortunately for this interpretation there is new evidence against any genetically consistent local stocks. Williams, Koehn, and Mitton (1973) showed that newly arrived elvers at some of the localities differed significantly from older residents. Unpublished additional findings show that elvers arriving in different years, or even a few weeks apart in the same year, may show differences comparable to those between localities hundreds of kilometers apart. Clearly the recruits to a given coastal locality are genetically heterogeneous. Variation among successive groups of recruits must result from differences in selection pressures in different water masses that deliver them to the coast.

Besides the abundant evidence on which Schmidt (1925) based his one-population theory, there are other supporting observations. Vladykov (1964) reported a north-south gradient in sex ratio; northern specimens are mainly female, southern ones male. There is no geographic variation in numbers of vertebrae or fin rays (Wenner, 1971), as is almost always found in widespread fishes (Jensen, 1944). This means not only that there is no strong locality-specific selection for meristic characters, but also that residents at widely separated localities must have had their early development in the same conditions. It is well known that different temperatures and salinities during early development produce different meristic characters in genetically similar stocks. The one-population theory can be considered more strongly supported now than ever before.

There are no other gene-frequency comparisons among age groups of fishes, but there are comparisons between size groups. Fujino and Kang (1968) found slight gene-frequency changes and marked decrease in heterozygote excess between juvenile and adult sizes of skipjack tuna. One-locus fitness differences of perhaps 20% for a fraction of the life cycle are indicated. Several less extreme size-group differences are reviewed by Ligny (1969) and are interpreted as evidence of mixing of different stocks. Differences in selection pressures on different age cohorts is an alternative explanation. If year classes are found to differ from each other but to be consistent through time, it would support my interpretation of high-mortality stages being the ones in which most of the selection takes place.

Studies of single-locus polymorphisms in low-fecundity organisms often indicate heterozygote advantage. Selection for sisyphean genotypes would maintain polymorphisms by frequency dependence and environmental heterogeneity. Heterozygote deficiency, rather than excess, is expected from this kind of selection, if there is ever any mixing of differently selected groups, even of different age cohorts. As expected, there are many convincing examples of heterozygote deficiency in marine fishes and very few of heterozygote excess (Ligny, 1969; Møller, 1969).

LOCAL VARIATION IN RELATION TO DISPERSAL IN OTHER ORGANISMS

Schopf and Gooch (1971) showed a cline in gene frequency in an ectoproct along the southern New England coast, with marked differences within 10 kilometers. They attribute the differences to an isolation-by-distance that is made possible by the young ectoproct's being planktonic only a few hours. This explanation is inadequate. A few hours is enough for tidal currents in this region to transport a larva 10 kilometers or more. I interpret the gene-frequency differences as entirely attributable to locality-specific and water-mass-specific selection on a genetically well mixed pool of larvae. The common mussel of this region has a larva that is planktonic for many days. Koehn and Mitton (1972) and Koehn, Turano, and Mitton (1973) found genetic differences between samples from intertidal levels mere meters apart. This should convince anyone that genetic differences can arise in the total absence of isolation in high-fecundity animals.

Benthic marine invertebrates with high ZZI are excellent material for testing the ideas presented here, and at least one student of these animals seems to share my belief that they are subject to intense selection, great variation in fitness, and the need for a strategy that can lead to genetically diverse offspring (Grassle, 1971).

There is considerable information on the population genetics of high- and medium-fecundity flowering plants, but it is often difficult to interpret because even order of magnitude estimates on the key factor of dispersal may be unavailable. Ehrlich and Raven (1969) incline towards conservative estimates for pollen dispersal, which they regard as normally on a scale of meters and only rarely kilometers. Faegri and van der Pijl (1966) believe in higher values, especially for wind pollination. They give as an "average maximum" a distance of 50 kilometers for wind pollinated trees, but state that pine pollen can be abundant hundreds of kilometers from the source. Levin, Kerster, and Niedzlek (1971) note that insect pollinators tend to continue in consistent directions and that this increases transport distances over those that would prevail with random movements between flowers.

Seed dispersal is also difficult to quantify. Larger wind-blown seeds of trees can commonly be seen to travel tens or hundreds of meters. Many weeds must have similar ranges of dispersal. Van der Pijl's (1969) review suggests that seed dispersal for several kilometers may be common in some species, but that a few meters is probably the rule for many.

The most convincing evidence of selection overcoming massive gene flow is in the recently studied examples of steep genetic clines where there are steep gradients in heavy-metal contamination of soils (Antonovics and Bradshaw, 1970; Antonovics, Bradshaw and Turner, 1971; McNeilly and Bradshaw, 1968; Cook, Lefebvre and McNeilly, 1972). Heavy-metal concentrations can change from negligible to severe in a few meters, and degree of resistance to the contaminant, by adult plants, shows a parallel steepness of gradient. A high heritability of resistance is experimentally demonstrated. A considerable gene flow by pollen and seed transport is apparent from comparing age groups. Plants on heavily contaminated soils bear large numbers of nonresistant or less resistant seeds. Likewise seedlings many meters from the contaminated area may be resistant, and adult plants on uncontaminated soil may produce a high proportion of resistant seeds. Pollen transport of 100 meters or more, in the direction of prevailing winds, is indicated by seed tests in wind-pollinated grasses. Nonresistant genotypes may be lethal with contamination levels that permit growth and successful reproduction by resistant types, and the resistant genotypes may be at a 50% competitive disadvantage on uncontaminated soil. Steep gradients in genetically determined chemical defense mechanisms in other plants are maintained by selective predation on seedlings (Crowford-Sidebotham, 1972).

The characters and genetic variation reviewed above are those that we would expect to be strongly selected, and may not be typical of fitness variation from other characters and gene loci. Also the soil contamination from mining wastes may be abnormally severe and the gradients unusually steep. Other evidence, however, less extreme but more natural, has suggested to botanists from at least the time of Turesson (1922) that steep clines are maintained by selection despite rapid gene flow. Turesson believed that most of the selection took place in high-mortality seedling stages. The evidence was recently reviewed by Stebbins (1970), who concludes that

The strong selective pressures exerted by the physical environment upon early developmental stages enable plant populations, even when normally outcrossed, to resist gene

89

flow and so to evolve adaptive constellations of genes to quite different habitats.

The ZZI of perennial forms with clonal multiplication may be extremely high and help to account for some of the intense selection on seedlings, but much of the evidence reviewed relates to annuals of only medium fecundity.

The only comparable work on a genetic cline in a high-fecundity tree is that of Barber and Jackson (1957). They found a steep genetic cline along an altitudinal gradient in a Eucalyptus. Quite different gene frequencies were found over horizontal distances of one kilometer, which they considered short in relation to normal dispersal. They concluded that

> . . . the selective coefficients per generation show almost the maximum possible change in value over less than 1500 feet in altitude.

Forest trees, as indicated by Hett (1971) and by Hett and Loucks (1968) are favorable material for demographic studies. This circumstance, and the possibility of new insights from studies of high-fecundity populations, make them desirable for genetic studies. New techniques such as isozyme typing make such work more feasible now than when Barber and Jackson did their research on Eucalyptus.

CHAPTER EIGHT

Natural Selection in High-Fecundity Populations: Evidence on Fertility

Viability and fertility are undoubtedly correlated components of fitness. Low-viability genotypes that reach maturity will probably have below average fertility, and low-fertility genotypes are less likely to survive to maturity. The purpose of this chapter is to show that there is great variability in fertility of different genotypes in nature. Fitness variation is therefore greater than merely its viability component discussed in the last chapter. Fertility variation has gotten less attention, both theoretical and observational, than viability variation. The only detailed consideration of fertility as a component of fitness is Bodmer's (1965), which indicates that the same general fitness models are usually applicable to both viability and fertility.

THE GENETIC COMPONENT OF FERTILITY VARIATION

There are both circumstantial evidence and theoretical reasons for believing that a considerable proportion of fertility variation in nature must be genetic. The circumstantial evidence comes mainly from experimental populations and analogy with genetic variation in other characters related to fitness. Theoretical reasons relate to the selective equivalence of any departures from optima, whether they are genetic or epigenetic in origin.

Agricultural research has produced a fund of information on differences between strains or varieties of crop plants (Baker, 1969; Jinks and Perkins, 1970; and references reviewed therein). Most agricultural varieties are more uniform than natural populations, but few are clonal. Comparison between varieties may overestimate, and within strains underestimate, variation in natural populations. There is normally a large genetic component (50% or more) of variability in such quantities as yield in grain, which would be closely related to fertility in nature. Variation in size, which is closely related to fertility in most plants, is known to have a large genetic component (Fripp and Caten, 1971).

There is also evidence for wild populations. Classic works on plant systematics (e.g., Turesson, 1922; Clausen and Heisey, 1958) were intended to estimate genetic differences between ecotypes, but incidentally provide information on variation within populations. For instance, Clausen and Heisey (pp. 19–28) showed mainfold genetic differences within ecotypes in size, vigor, and other characters related to fertility.

Of various measurable characters, only viability and fertility, measured in nature, can be related linearly to fitness. Measures of adaptive performance, such as fleetness or stress resistance, must be monotonically related to fitness, but a simple linear relation between fitness and the investigator's measurements is unlikely. Many easily measured metrical characters must have optima near the mean of the distribution, and fitness would decline with increasing steepness in both directions from the optimum. O'Donald (1971) and others advocate a general rule that fitness should decrease as the squared deviation from the optimum. Regardless of the exact relationship, it seems safe to assume that if a character is more variable it will cause more variation in fitness than if it were less variable. It can also be assumed that if a large fraction of the observed variability of a character is genetic, a large fraction of the variability in fitness caused by that character must be genetic.

Kerfoot (1969) showed from maternal-foetal comparisons

that about 65% of scale-count variance in a snake must be genetic. To whatever extent fitness is correlated to scale counts, a considerable part of its variation must be genetic. Kerfoot's technique really measures heritability, which must be less than the total genotypic determination of a character, and must therefore give minimum estimates. Maternal-foetal comparisons in natural populations of eel-pout (Schmidt, 1917; Ege, 1942) and artificial stocks of guppies (Schmidt, 1919) also show heritabilities in the range of 50% for various meristic characters.

The evidence is hardly as satisfactory as might be wished. All that can be said is that (1) there is considerable genetic variability in fertility of cultivated plants, and no reason to assume that they are markedly different in this respect from their wild progenitors; (2) there is no reason to assume that fertility variation is markedly different from other quantitative characters in natural populations, which generally prove to be largely genetic in origin; and (3) many characters closely correlated with fertility, such as size in many organisms, are known to have great variation.

Theoretical considerations make it difficult to see how any character related to fitness could fail to be subject to both genetic and environmental influences. Suppose a character optimum is χ_o and an individual has the suboptimal value $\chi_o + \Delta\chi$. It makes no difference to selection whether the departure from the ideal results from having a genotype that specifies $\chi_o + \Delta\chi$, or one that specified χ_o but permitted environmental influences to alter the specified value by $\Delta\chi$. Selection is equally concerned to eliminate wrong-value genotypes and reduced-canalization genotypes. So as long as selection is a dominant force in determining genotypes that prevail, it seems inevitable that departures from optima will have both a major genetic and a major environmental component.

If we knew that reduced fitness from wrong-value genotypes tends to arise by mutation and recombination to the same extent as from reduced-canalization genotypes, we could infer that the genetic and epigenetic loads must always be equal. Equal production of the two kinds of mutation would follow from shifting the argument from mutations that affect phenotypes to mutations that affect mutations. A gene that permits increased production of either kind of reduced-fitness mutation would be equally selected against. The relationships of genetic to epigenetic load ought to get some attention, both theoretical and experimental.

The remainder of the chapter is based on the assumption that a considerable fraction, perhaps about half, of the fertility variance in natural populations is of genetic origin. It should be borne in mind that the observed variation is reduced to the degree that fertility and viability are correlated. We only measure fertility in organisms that have, in fact, survived to sexual maturity. For any observed fertility distribution, we can assume that many abnormally low values, that theoretically ought to be there, have been culled out because low-fertility genotypes are often low-viability genotypes.

FERTILITY VARIATION IN HIGHER PLANTS

As noted in the preceding chapter, size variation in an age cohort of higher plants is often enormously variable. The close dependence of fertility on size must mean that fertility is enormously variable. The only major source of quantitative data on plant fertility is Salisbury's (1942) compilation of counts of fruits per plant and seeds per fruit of mostly herbaceous plants of Britain. They are mainly of medium fecundity (10^3 to 10^6 as suggested in Chapter 6). Seed counts of large woody plants would have been more germane, and the data are unsatisfactory in other respects. Those for a single species may lump individuals of different ages and different localities. From Salisbury's many examples I have chosen only those that can have seed output in at least the tens of thousands per year, that are at least mainly monocarpic (to assure that the data relate to a single cohort), that come from a single locality, and that include counts on at least a hundred specimens. The data support the conclusion that fertility variation is enormous in the populations studied (Figure 12). The top 10% usually has several times the fertility of the bottom 10%.

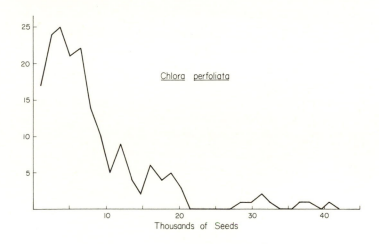

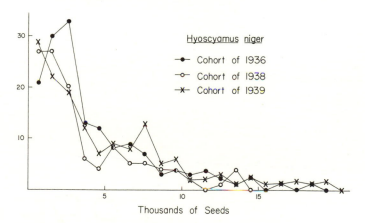

FIGURE 12. Seed fertility distributions of monocarpic plant populations, from Salisbury (1942).

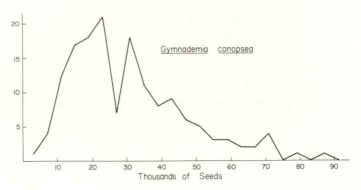

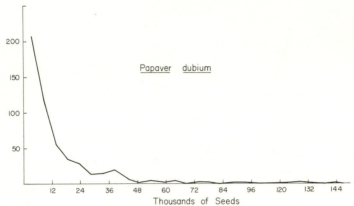

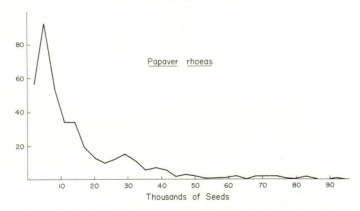

FIGURE 12. Continued.

96

A positive skew is apparent for every distribution. Salisbury admitted to under-representing the smaller and more easily overlooked members of a population. If even the smallest specimens had been represented equally with the largest, the distributions would be even more skewed.

A recent study of Papaver by Harper (1966) indicates an even more variable and skewed distribution than those found by Salisbury. It is to be hoped that other such work will be done, preferably on trees, and with counts for different age cohorts in single stands and of single cohorts over several years.

FERTILITY VARIATION IN FISHES

Size variability in fishes implies great fertility variability, because female fertility is closely related to size. Bagenal (1966) concluded from voluminous data, on plaice and other prolific species, that age has no important influence on fertility, apart from its effect on size. Since the fertility variability of females is in addition to (and partly correlated with) viability variation, the assumption of great variation in fitness is thereby supported. Chapter 11 indicates that fertility variation in males is greater than in females in most species.

Actual data on fertility in relation to age are less common than on fertility and size, but such information is gathered from time to time (Figure 13). There are undoubtedly other relevant studies, but those presented should be sufficient. They were chosen solely on the basis of availability of publications and the adequacy and relevance of information. I required at least twenty fertility values per cohort, which were either measured on scatter diagrams or taken directly from tables.

Unfortunately there is no assurance that the samples represent single populations in the Mendelian and ecological sense. They are merely fishes caught in a limited area with standard commercial gear, which may be biased in favor of larger specimens. The lamprey data came from several rivers, but all had

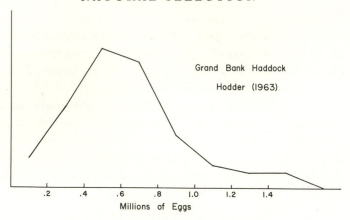

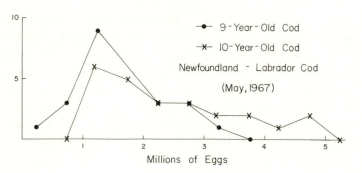

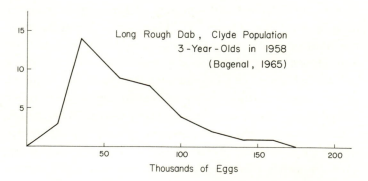

FIGURE 13. Single-season egg production in fishes.

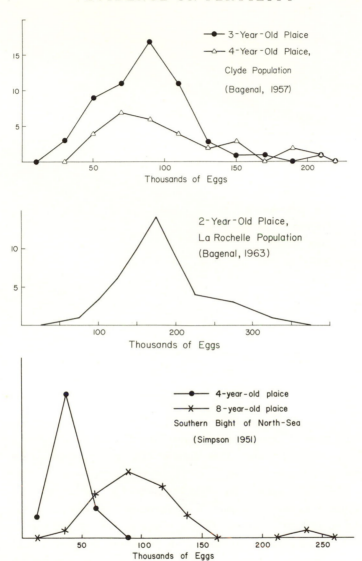

FIGURE 13. Continued.

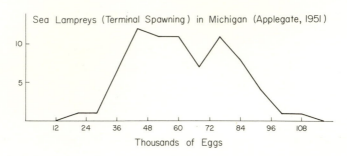

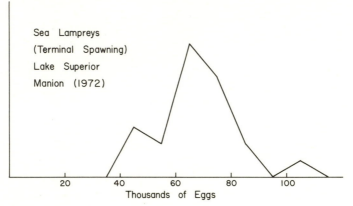

FIGURE 13. Continued.

spent most of their lives in a single lake, and all are descended from a few founders that entered the upper Great Lakes through canals.

The distributions all show considerable variability and conform to the lognormal distribution as well as can be expected of small samples. This is especially true of the species with the higher fecundities. Only with the lamprey would there seem to be any doubt. This species matures at a certain size, spawns, and dies. Spawning groups may therefore be of variable age, but not of very variable size and fecundity. The bimodality of Applegate's data must reflect his stated aim of giving adequate representation to both the largest and smallest specimens.

There seems to be a tendency for variability to increase with age. An extreme example is the trout studied by Wydoski and Cooper (1966), in which differences between upper and lower ends of the distribution increased from about two-fold to eight-fold in three years. The gadid and flatfish populations in Figure 13 are all heavily fished, except possibly the long rough dab. The greater life expectancies in unexploited populations would increase fitness variability measured over whole lifetimes. It follows from principles operative in the evolution of senescence (Hamilton, 1966; Emlen, 1973) that fertility differences and other aspects of fitness variation should increase with age. The effectiveness of selection should decline with increasing age to exactly the same extent for genetic and epigenetic load.

Data summarized above indicate that egg or seed fertility in nature is widely variable. The distributions generally show better conformity to lognormal than to symmetrical or truncate models and support the assumption of multiplicative interactions among genetic and environmental contributions to this important component of fitness. In conjunction with theoretical considerations (Chapters 5 and 6) and data on viability (Chapter 7) they support the inference that fitness in high-fecundity populations can be approximated by high-variance lognormal distributions and that only a few sisyphean genotypes make an effective contribution to the next generation. The cost of meiosis is thereby a minor part of a commonly manifold difference in fitness among genotypes. Its parthenogenetic avoidance would not be worth the sacrifice of benefits from genetically diverse progenies, as indicated in the earlier chapters.

CHAPTER NINE

Derived Low-Fecundity Populations

Chapters 4 and 5 proposed ways in which sexual reproduction might entirely replace asexual in some evolving lines. Even on the improbable assumption that all the reasoning in those chapters is correct, the question remains: correct for what organisms? Exactly what kinds and degrees of dispersal are necessary, and how heterogeneous must environments be, and what is the minimum necessary fecundity? I suggested on page 43 that the fecundity problem might be approached by a study of plant or animal taxa in which closely related forms may differ as to the presence or absence of asexual reproduction. When major taxonomic groups all share a certain feature, it is unlikely that the feature has the same adaptive significance throughout the group. It may even be maladaptive for the majority. The crossing of digestive and respiratory systems of vertebrates is a good example. Similarly, asexual reproduction is unknown among the mollusks. It seems most unlikely that this could be a uniformly advantageous deficiency.

While it may be difficult to identify any species for which it can be confidently asserted that exclusively sexual reproduction is selectively advantageous, it is easy to identify some in which it must be disadvantageous. All organisms with really low ZZI such as mammals, birds, and many insects have populations in which sexual reproduction must be consistently selected against. Their present exclusive reliance on sexual reproduction must be ascribed to inheritance from a high-fecundity ancestor in which the complete replacement of asexual with sexual reproduction was the evolutionary equilibrium. If and when any form of asexual reproduction becomes feasible in higher vertebrates, it completely replaces sexual. So in these forms sexuality is a maladaptive feature, dating from a piscine or even protochordate ancestor, for which they lack the preadaptations for ridding themselves.

OBSTACLES TO THE EVOLUTION OF ASEXUAL REPRODUCTION IN HIGHER ORGANISMS

Higher animals obviously lack the potential for vegetative multiplication by adults, and the same may be true for many single-stem plants. Still-plastic mammalian embryos may reproduce by fission as an occasional abnormality. It would be adaptive for the embryo but certainly not for the mother. She has already paid the price of meiosis. Genetic homogeneity among her offspring would give her some disadvantages of asexual reproduction without its primary benefit. The conflict of generations seems generally to be resolved in favor of the mother, for reasons that are not immediately obvious. A successfully divided embryo may have nearly doubled its fitness. With litter size n, the increased cost to the mother is only $1/n$, very small in species with large litters, but polyembryony is normal only in armadillos (Tyler, 1967).

A bird or reptile must stock each egg with food reserves appropriate for getting one young to the hatching stage. I assume that two embryos utilizing this reserve would produce two hatchlings with a combined reproductive value less than that of a single normal hatchling. Comparable considerations must apply also to insects.

It is also easy to imagine obstacles to the evolution of parthenogenesis in any organism. Evolution of a haploid parthenogenetic population from a sexual diploid would encounter the problem of full expression of all deleterious recessives. Reversion to diploidy by sexual reproduction would be expected at first opportunity (see Dodson, 1953; Crow and Kimura, 1965; and Maynard Smith, 1971B, for advantages of diploidy). The trouble with diploid parthenogenesis is that it can become feasible only if several kinds of change occur simultaneously and completely. Meiosis must be entirely suppressed or cancelled by resorption of the first polar body. Use of the second polar body would be equivalent to selfing, with maximum inbreeding depression. Partial suppression of meiosis would produce chromosome abnormalities that would probably be lethal. Polyploidy via hybridization is the most likely shortcut to suppressed meiosis. As a rule, exclusively parthenogenetic animals have an altered ploidy and other evidence of hybrid origin.

Even with meiosis suppressed, other problems remain. The diploid or polyploid egg must resist fertilization. Otherwise an inviable or sterile individual of odd ploidy would be produced. The egg must initiate and continue normal development without the signal of fertilization, and without the centriole and mitochondria supplied by the sperm. Both the sudden and gradual acquisition of these characters would seem unlikely in an organism historically committed to exclusively sexual reproduction. These problems apply mainly to animals. In plants, apomictic seed production can follow replacement of the haploid egg by a diploid vegetative nucleus (Clausen, 1954; Haskell, 1966). This may account for the commonness of facultative parthenogenetic development in higher plants (Gustafsson, 1946; Stebbins, 1970).

Experiments on induced parthenogenesis in normally sexual species are often reported without any attention to the ploidy of the uniparental young. When this problem is investigated, the individuals often turn out to be diploid (Beatty, 1957). There is seldom any evidence as to whether diploidy resulted from the absence of meiosis or from secondary restoration that may be equivalent to inbreeding. Likewise there is seldom any indication as to whether the experimental stimuli caused the diploidy or selected for already diploid eggs. Unfertilized eggs occasionally develop spontaneously, and where this can occur, selection can increase its frequency. Olsen (1965) produced a strain of turkeys in which many unfertilized eggs showed some development, and one in several hundred reached maturity. No observations of ploidy were given, but since the birds produced were all males, some alteration of the maternal genotype must have occurred.

The situation is clearer in Drosophila, as a result of Carson's (1967) experiments. He selected for parthenogenesis in Drosophila and increased its frequency from .001 to .064. They seemingly developed from unreduced eggs, because all parthenogenetic offspring were diploid, vigorous, fertile, and female. It may be too much to expect that such eggs would also resist fertilization. I suggest that, if normally inseminated, the partly parthenogenetic females would produce about 6% inviable triploids.

PARTHENOGENESIS IN HIGHER ORGANISMS

Despite the obstacles, an apparently asexual parthenogenesis is occasionally evolved by vertebrates. There are parthenogenetic fishes, amphibians, and reptiles, and there are fishes that are genetically parthenogenetic, but with eggs that require insemination by males of other species (*gynogenesis,* see White, 1970). It would appear that as soon as parthenogenesis is possible in any vertebrate population, sexual reproduction is "immediately" lost. There are no heterogonic vertebrates, with periodic or facultative sexuality. This would require environmental cues to synchronize sexual reproduction at the optimal moment, as envisioned in the Aphid-Rotifer Model. In organisms with several generations per year, this can be done with the seasonal cycle. There is no longer cycle to provide reliable cues for more slowly maturing forms. Some fishes and a few other vertebrates, such as the smaller rodents, can produce several generations per year. Conceivably among the fishes the introduction of parthenogenesis could result in the evolution of a heterogonic life cycle with asexual and sexual reproduction in evolutionary equilibrium. No rodent would have a high enough fecundity to achieve the requisite ZZI in a single year. The fact that parthenogenesis or its equivalent, if found in a vertebrate population, has always replaced sexual reproduction entirely, is decisive evidence of the maladaptive nature of sexuality in these organisms.

Even among invertebrates, it would appear that the initiation of parthenogenesis normally results in the loss of sexuality. Secondarily heterogonic life cycles are characteristic of some major groups, such as aphids and rotifers, but the taxonomic distribution of such life cycles suggests that they have only evolved a few times. Among weevils and flies, parthenogenesis has evolved independently several times, and in each case sexual reproduction disappeared (White, 1970).

The facultative apomictic seed production of many plants must be in equilibrium with sexual seed production, as envisioned in the Aphid-Rotifer Model. Obligate apomictic seed production is less common and usually associated with sexually sterile hybrids and polyploids (Asker, 1970, 1971, and earlier papers in this series). It results from the inability of plants with abnormal chromosome complements to proceed with a normal meiosis. Similar conditions arise in the ferns and give rise to ameiotic sporogenesis and parthenogenetic egg development (Evans, 1969). I would attribute the scarcity of asexual spore production by ferns to their enormous spore fecundity and conformity to the Elm-Oyster Model.

UNIPARENTAL SEXUAL REPRODUCTION

Various modes of reproduction that might seem intermediate between asexual and sexual are discussed in Chapter 10. For the present discussion, two highly adaptive reproductive modes can be recognized, the perfectly asexual and the outcrossed sexual. With each we can associate a complete set of

106

selective advantages. This is not true of any compromise between the two. Self-fertilization, endomeiosis, restoration of diploidy with a polar body, etc., may offer no advantages over asexual reproduction, and may offer serious disadvantages from loss of genetic material or inbreeding depression. I interpret the prevalence of these uniparental sexual processes as indications that perfectly asexual egg or seed production would have been evolved, given the requisite preadaptations. The clearest case is the gynogenetic fishes. Their reproduction is genetically asexual, but they retain the need for the stimulus of insemination to initiate development. If development without this stimulus could be evolved, these all-female fishes would be free of the necessity of living only in areas inhabited by related species that they can beguile into a genetically unproductive contribution of gametes.

Somewhat analogous are those plants that have achieved a genetically asexual seed production, but are still burdened with the necessity of endosperm fertilization (Haskell, 1966; Nygren, 1966). In some flies meiosis takes place but diploidy is restored by resorbing the second polar body. A considerable proportion of lethal homozygotes are thereby produced (Stalker, 1956), but their loss is apparently worth the partial avoidance of the cost of meiosis.

Self-fertilization avoids the cost of meiosis as a genome-wide average. For every gene that is abandoned another is doubled. The rarity of habitual self-fertilization indicates that the resulting inbreeding depression is so severe as to offset the 50% advantage of avoiding the cost of meiosis, at least in the majority of plants and animals. Most plants outcross if they can and self-fertilize as a second choice. In populations that must often resort to the second choice, genomes tolerant to high levels of homozygosity will evolve. Inbreeding depression may be mitigated to the point at which the cost of meiosis will be a greater loss. Thereafter the plant will preferentially self-fertilize, and outcrossing will be a rare abnormality.

Flukes and tapeworms (Noble and Noble, 1961) and fresh-water snails (Hyman, 1967) can also self-fertilize. Like most plants, they do so only if isolation prevents outcrossing. So in these forms also, inbreeding must cost more than meiosis, although under ideal conditions the snails can be propagated for many generations by self-fertilization.

LINKAGE AND RELATED PHENOMENA

Given that reproduction will be sexual and outcrossed, and the cost of meiosis paid in full, there remains a wide range of possibilities. One hypothetical extreme would be to have every gene locus segregate independently of every other. The other extreme is realized in hybridogenetic fishes (Schultz, 1971), in which no recombination within a genome ever takes place. The paternal genome is discarded intact in oogenesis, and the original maternal genome transmitted indefinitely, with only mutation as a possible source of change.

The presumably typical situation is for chromosome pairs to segregate independently, but for linkage within chromosomes to keep some genes associated for a number of generations. I assume that observed chromosome numbers and crossover rates reflect the optimum compromise between maximizing whatever benefits there are in recombination, and minimizing recombinational load. Tighter linkage must reduce recombinational load, but it does nothing to alleviate the cost of meiosis. It is therefore of limited interest on the problem of when recombination is worth the cost of meiosis and when it is not. It does regulate the rate of recombination between some gene loci when reproduction is sexual, and if we understood this regulation, we would surely have at least a partial grasp of the significance of recombination in general.

The current evolutionary literature shows a keen awareness of recombinational load and of factors that reduce it by reducing recombination. These factors are presumably adaptive and can be selected for, hence Turner's (1967c) query, "Why Does the Genotype Not Congeal?" If all genes were on a single pair of chromosomes, with no crossing over, the presumably desirable condition of zero multi-locus recombinational load would be achieved. Some forces must be opposing this development, and Turner (1967a; 1967b; 1970; 1971) suggests that these forces lie in side effects of changing chromosomal structure, in complex and poorly understood gene interactions, and in environmental uncertainty. I would emphasize this last factor.

In natural populations we expect selection for canalization to make adaptive characters dominant to maladaptive. The limited information (Kidwell, 1972) indicates that low rates of recombination are dominant to high. This suggests that recombination is something that selection has been minimizing, but response-to-selection experiments are equivocal on this same point. If selection in nature is generally for low rates, it should be difficult to reduce them further by artificial selection, but easy to increase them. This is, in fact, the pattern sometimes observed, but at other times it is as easy to select for decrease as increase (Chinnici, 1971). Different organisms can give different results, as can different gene loci in the same organism.

If the views expressed in this chapter are correct, higher vertebrate and other low-fecundity populations are all reproductively maladapted, in the sense that sexual reproduction would disappear if an alternative asexual process were introduced. Perhaps at any given evolutionary moment a fair proportion of these organisms are near the point of discovering a feasible way of reproducing asexually, or by a partly equivalent degenerate sexuality. It may be that in any period of a few million years a sizeable number of them do make the discovery. On the usual vew of the significance of sex in evolution (Chapters 12 and 13) they may thereby commit a phylogenetic suicide.

They would be replaced by offshoots from near relatives that have not yet made the discovery. This kind of bias in taxonomic extinction, which should not be confused with group selection as usually understood (Chapter 13) could keep exclusively asexual species in low frequency in the biota, despite a frequent loss of sexuality.

CHAPTER TEN

Patterns of Sexuality

The presence of meiotic oogenesis in higher animals and plants is the major challenge in sexual reproduction, but clearly there are many others. This chapter sketchily considers some other problems, especially those related to the main one of the cost of meiosis. Related issues on the roles of the sexes in dioecious animals are reserved for Chapter 11.

PRIMITIVE SEXUALITY

Asexual reproduction is any process that produces offspring in such a way that they precisely inherit the parental genotype. It requires the prior evolution of exact mechanisms for preserving individuality and genotypic integrity. Bacterial transduction and transformation are symptoms of the imperfection of these mechanisms in organisms that lack well-organized nuclei and mitotic machinery. This sort of recombination must have been frequent in early, really primitive organisms. Although not "meiotically diversified" (Table 1, p. 4) their reproduction must have been, in effect, sexual, and much evolutionary advance had to occur before genetically uniform cloning became possible. These ideas have been expressed before (Boydan, 1954; Ehrensvärd, 1962; Parker, Baker, and Smith, 1972), but historical priority is often assumed for asexual reproduction.

As long as a bacterium or protist cell is living under conditions that provide the necessities for increase in number, it is likely that the same genotype, one cell-generation later, will enjoy the same promising conditions, but if there is difficulty in mere maintenance, prospects are unfavorable. A gene that maintains precision of cloning when conditions are optimistic, but permits recombination when they are not, will be favorably selected. A change from favorable to unfavorable conditions for a given genotype corresponds to a loss of heritability of fitness, and sexual reproduction should continue to occur when this heritability is at its minimum in the life cycle. Arguments of the Aphid-Rotifer Model apply, except that (as explained below) it is only necessary for advantages of recombination to overcome recombinational load, not the two-fold cost of meiosis. As soon as cloning was possible in the history of life, sexual reproduction would be confined to times of change or stress.

A primitive living system must occasionally have lost its hold on an information-bearing molecule, which would escape into the medium and become available for incorporation into another system. Parker, Baker, and Smith (1972) draw the analogy between this kind of recombination and phage infection. A gene with this mode of inheritance would have the disadvantage of possibly being used as a substrate or destroyed abiotically, rather than being preserved for its information content. If a gene favored the preservation of the genome except following fusion with a somewhat different individual, its likelihood of loss would be minimized. Maynard Smith (1966) points to the possibility of this primitive syngamy not being very different from phagocytosis at an early stage. Genetic reorganization would come to be synchronized with syngamy, and a primitive haploid prokaryote with a momentarily diploid zygote would have appeared.

I am indebted to F. C. Warburton for any clarity there may be in the following ideas on meiotic costs in isogametic organisms. If a zygote immediately undergoes meiosis to form genetically distinct haploid clones, all genes that entered with the gametes are retained, and there is no cost of meiosis. It might be argued that each of the clones has only half of the zygote genotype, and that there is a 50% cost of not remaining diploid. Such an argument assumes that it is just as feasible to produce four diploid as four haploid cells from the first two divisions of the zygote, a questionable assumption for one-celled organisms. A large multicellular organism can double the DNA of an egg nucleus from reserves in yolk or parental soma at trivial cost. The gamete of a unicellular organism is the whole organism, with no external somatic resources, and the arguments on haploid and diploid eggs being about equally expensive to the parent (pages 9–10) do not apply. Its zygote has the options only of making two large diploid, or four small haploid cells from material in the zygote. The concept of cost of meiosis applies only weakly or not at all to isogamous organisms.

The artfully fashioned models of Parker, Baker, and Smith (1972) provide welcome insights into the origin of anisogamy. They show that, given plausible assumptions on how survival prospects may be related to zygote size, there will be disruptive selection for gamete size. It will be advantageous to make gametes large enough to give an adequate-size zygote, even if fusion with an undersize gamete occurs. Given that such large gametes are available, there will also be selection to exploit the situation by making larger numbers of small gametes and thus increase the number of large ones found and fertilized. There would be selection for the ability to discriminate against small gametes so as to maximize cytoplasmic resources available to the zygote, but this selection would be most intense for the small gametes. Their genetic survival absolutely requires a large partner, which would get a merely quantitative benefit from fusing with another large gamete. For the same reason, there would be intense selection for the small gametes to be able to overcome the mechanisms of discrimination used by the larger ones. This primeval conflict between the sexes was resolved in favor of the males, because of more intense selection for male functions. Once the issue was decided, macrogametes abandoned the attempt to unite with other macro-

113

gametes and reject sperm. The theory is the same, whether the two kinds of gamete-making strategy are employed by a hermaphrodite or two dioecious organisms. One sex may be regarded as parasitizing the parental investment of the other. In hermaphrodites the parasitism is mutual.

Once anisogamy is evolved, the only way to escape genetic parasitism by sperm is to abandon syngamy and become parthenogenetic. Any female that can accomplish this, in an otherwise exclusively sexual population, thereby doubles her fitness. One minor problem that remains is the origin of polarbody formation. The result would be the same whether n eggs with polar bodies were formed from n oogonia or without polar bodies from $n/4$ oogonia. I suggest that it provides better control over events in the egg for it to remain diploid as long as possible, until after the egg is stocked with yolk reserves. Once the cytoplasm is really bulky, symmetrical fission becomes mechanically difficult but the budding off of minute polar bodies very easy.

CLASSIFICATION OF MODES OF REPRODUCTION

Mahendra and Sharma (1955) proposed that reproduction be divided into *meiotic* and *ameiotic,* which would be equivalent to sexual and asexual as used here. There is a problem in the possibility that meiosis could occur and reproduction still be, in effect, asexual, as with "fertilization" by the first polar body. Also the presence or absence of a cytologically recognizable meiosis is less important than the genetic distinction of whether recombination is present or absent. *Recombinational* and *nonrecombinational* would express exactly the important distinction. The more facile terms sexual and asexual can do the same, and have been used in this sense by recent writers on the evolutionary significance of recombination.

In asexual reproduction a major distinction would be

whether it relates to growth processes (vegetative reproduction) or utilizes structures more primitively used in sexual reproduction (parthenogenetic eggs, apomictic seeds, mitotically produced spores). It is also important to distinguish primitive from derived asexuality. The derived would include all instances of asexual reproduction in organisms that had an exclusively sexual ancestor. Ameiotic parthenogenesis is always derived. Some instances of vegetative reproduction in various plants and worms may also be derived, as is, certainly, the polyembryony of the armadillo.

Most sexual reproduction can aptly be termed *euphrasic*. It requires production of haploid gametes that must unite in pairs to form an at least momentarily diploid zygote. The gametes that unite must come from different individuals so that the zygote generation is genetically diverse. Whether there is isogamy or anisogamy, and whether the parents are monoecious or dioecious are secondary considerations. The most important of the subordinate distinctions would be whether macrogametic haploidy results from a self-imposed genetic cost, or from nutritional considerations.

Finally there is what may be called degenerate sexuality, of which self-fertilization by hermaphrodites would be the commonest form. Another possibility would be the restoration of diploidy in a haploid egg by any of the methods discussed by Crew (1965). Asher (1970) and Maslin (1968) have analyzed the theoretical consequences of various modifications of the meiotic process prior to parthenogenesis. It is characteristic of all forms of degenerate sexuality that the population tends towards complete individual homozygosity (but often with high levels of polymorphism in the population). Asher showed that the tendency towards homozygosity can be opposed by selection so as to give a fairly high equilibrium level of heterozygosity. Disadvantages of degenerate sexuality were discussed on pages 106–108.

Bisexual is sometimes used to describe populations of male

115

and female individuals, and sometimes the opposite, of each individual being of both sexes. For this reason Tomlinson (1968) urged that the term be avoided altogether. He also pointed out that *gonochoristic* and *dioecious* seem to mean the same thing. Reproduction may also be divided into *uniparental* and *biparental* (Mayr, 1963). This avoids the difficult problem of carefully distinguishing such things as parthenogenesis and self-fertilization, but unfortunately, these difficult distinctions are the theoretically important ones. The genetic consequences and evolutionary significance of self-fertilization and ameiotic parthenogenesis are entirely different, and ignoring the distinction will obscure some important issues.

In discussing the Aphid-Rotifer Model I made the customary assumption that parthenogenetic reproduction in the model organisms is in fact clonal. This has only recently been clearly established for rotifers (Birky and Gilbert, 1971) and for Daphnia (Herbert and Ward, 1972). Aphid oogenesis resembles that of Daphnia in the formation of a single polar body which remains outside the egg, and at no stage is the egg visibly haploid. Unfortunately there are no genetic data to support the clonal nature of aphid parthenogenesis, and the point remains inconclusive (Bacci, 1965).

The suggested classification of modes of reproduction (outlined below) is designed for discussions of the adaptive significance of sexuality. For other purposes, of course, other classifications of reproductive processes may be more appropriate.

Sexual reproduction

Primordial. Offspring genotypes altered because mechanisms of genotypic maintenance are poorly developed. Modern analogies: bacterial transduction and transformation.

Euphrasic
Conservative (isogamy of haploid protists, with no polar bodies formed). Examples: sexual reproduction in most protozoans and simple algae.

Costly (50% genetic loss in oogenesis). Examples: sexual reproduction in diatoms(?), elms, man.

116

Degenerate
Selfing by hermaphrodites. Examples: Mendel's peas, some fishes.

Parthenogenesis, whenever the process fails to preserve the maternal genotype. Examples: artificially induced development of haploid eggs.

Others?

Asexual reproduction

Primitive (always fissile or vegetative). Examples: mitotic fission in protists; vegetative reproduction of sponges.

Derived
Vegetative. Examples: polyembryony of higher animals; probably some growth-related spread of higher plants; probably fission in some worm groups.

Amictogametic. Examples: any development from ova, spores, etc that preserve the maternal genotype.

COMPARATIVE SEXUALITY OF HIGHER ORGANISMS

Evolutionary biology is devoted to explaining diversity, and life cycles are as diverse as the structures seen by the comparative anatomist. Consider that a single class of animals can include the life cycles of aphid, honey bee, 17-year cicada, and mosquito. Perhaps the first century of Darwinism can be excused for not often giving explicit recognition to the idea that the organization of a life cycle is a product of natural selection. Today I sense an increased awareness of the great wealth of challenges in the evolution of life cycles, and some important pioneer work has been accomplished (Gadgil and Bossert, 1970; Hamilton, 1966, 1967; Lloyd and Dybas, 1966A, 1966B; Parker, Baker, and Smith, 1972).

For organisms that practice euphrasic sexual reproduction, we have only made a bare beginning at explaining the diversity to be seen. Mather (1940) was the first to theorize, in a manner acceptable by modern standards, on the phylogenetic distribution of hermaphroditism in the animal kingdom. He argued that dioecy is primarily a mechanism of outbreeding, and hermaphroditism a compensation for locomotor deficiencies of parents. The most important recent contribution on animal hermaphroditism is Ghiselin's (1969) review, which proposes several evolutionary explanations for the observed distributions. This work and Maynard Smith's (1971B) analysis of hermaphroditism in relation to cost of meiosis are discussed below.

Bacci (1965) reviewed parthenogenesis in relation to sexual reproduction and the cytological cycles of heterogonic animals. Parthenogenesis and gynogenesis in the vertebrates are reviewed by Beatty (1967) and Uzzell (1970), and insect parthenogenesis by Suomalainen (1969). White (1970) discusses parthenogenesis in both vertebrates and insects, and Oliver (1971) in arachnids. McGilvray (1972) reported life-cycle variation within a species of aphid. In some areas it shows the usual heterogonic life cycle. In others it is entirely parthenogenetic, and in still others fertile males are produced, but no mictic females.

Hawes (1963) discussed the distribution of asexual and sexual reproduction in protozoa and concluded that absence of sexuality is a derived condition evolved several times independently. The protozoa offer many complex problems in comparative reproduction. One need not go beyond the genus Paramecium for a wealth of unexplained cyto-sexual phenomena. Incompatibility systems of certain fungi offer challenging evolutionary puzzles (Raper, 1966). All the works listed are mainly descriptive reviews, but provide mines of information for anyone interested in generalizing on ways in which sexual reproduction is adaptively modified in relation to other aspects of a life cycle.

Lewis' (1942) classic work on comparative sexuality of higher plants marshalled evidence that a frequent trend is from (1) having both sexes in each flower to (2) having male flowers and female flowers on the same plant to (3) having separate male and female plants. The taxonomic distribution of these characters suggests that the final stage, if well established, is irreversible, but that these dioecious plants have a greater rate of extinction than the other two. On the forces of selection operating on these characters, Lewis generally follows Mather (1940) and partly anticipates Ghiselin's (1969) Gene Dispersal Model discussed below. Lewis' work is full of theoretical suggestions and ideas for testing them with comparative data.

Higher plants may be especially favorable material for testing theories on the evolution of sexual processes. Vegetative multiplication, mictic and apomictic seed, various forms of hermaphroditism and self-incompatability systems may or may not be present, and may or may not be related to variation in ploidy (Asker, 1971; Ornduff, 1971). That great variation can occur within a genus shows that plants are often preadapted for changes in these characters and may respond rapidly to selection for modification of reproductive strategy. Cytological mechanisms of higher plants are usually more easily studied than those of animals and show great diversity (Evans, 1969; Wet, 1968).

This section does little more than indicate that one can recognize a field of comparative anatomy of life cycles, that various modifications of sexual reproduction form a key element in the organization of any life cycle, that a fair mass of relevant data has been accumulated, and that some rudiments of conceptual systematization have been accomplished. The main work of providing a workable theoretical structure for understanding the enormous diversity of life cycles remains to be done.

CONTROVERSIAL ASPECTS OF HERMAPHRODITISM

An important principle that emerged forty years ago with R. A. Fisher is that there can be no average advantage of one sex over the other in dioecious organisms. Various implications, such as rates of parental investment in sons and daughters, and the stage at which males and females should be equally numerous, have been worked out by more recent workers. An implication that has not gotten much attention is that male and female functions must be equally expensive. Hermaphroditic plants, for example, must be spending resources equally on pollen production and distribution and on ovule and seed production. If it were otherwise, if, for instance, it were less of a sacrifice to pollenate other plants than to be pollenated and nourish fertilized embryos into mature seeds and associated structures, a plant that increased its pollen production and reduced that of seeds would have a selective advantage. The population would evolve higher pollen and lower seed production, until it was spending equally on both.

The expense of pollen production should make hermaphroditic plants less productive of seeds than female ones of the same species. This expectation is well confirmed (Lewis and Crowe, 1956). The amount of depression in seed fertility expected in hermaphrodites depends on possible correlations between hermaphroditism and size or vigor, on the population sex ratio, and on frequency of selfing (Ross and Shaw, 1971; Valdeyron, Dommee, and Valdeyron, 1973).

In a completely hermaprhoditic population, not every individual need spend equally on male and female functions, but the population as a whole must. This conclusion should apply to all outcrossed simultaneous hermaphrodites, regardless of physiological details. In this I follow Parker, Baker, and Smith (1972) and disagree with Maynard Smith's (1971b) reasoning that hermaphroditic animals with efficient (internal) fertilization will spend little on male functions and thereby largely avoid the cost of meiosis. If the male functions were really cheaper, the individual that fertilized more and was fertilized less would be at an advantage and the population would evolve a greater emphasis on male functions. Only when selfing is the rule would the cost of meiosis be largely avoided.

With selfing there is no competition between individuals for fertilizable eggs. Selection should maximize functional efficiency in fertilization and not favor wasteful mechanisms of competition. I assume that a normally self-fertilizing plant or animal spends more on female than on male functions.

This conclusion that male and female functions are equally expensive in outcrossed hermaphrodites, even when fertilization is internal, can be supported or refuted by studies on reproductive behavior and physiology in animals that meet the specifications, such as oligochaets and pulmonates. General works on the anatomy of these organisms illustrate male reproductive organs as complex systems of comparable bulk to the female. Sexual contact in oligochaets is a mechanically complex process that takes from many minutes to several hours, with larger species requiring longer time (Stephenson, 1930). There is nothing to suggest that male functions are especially cheap. The period of contact must include both courtship and copulation, with courtship serving mainly to enable each participant to gather information on both the male and female suitability of the partner.

The expense of maleness is especially convincing in the pulmonates. Male reproductive systems are bulky arrays of specialized organs, and courtship is an elaborate affair lasting as much as three hours. The self-fertilization that may occur in isolated individuals takes place in the gonad and makes no use of the accessory male organs, which are designed for courtship and copulation. These would appear to be as wastefully competitive in these hermaphrodites as in dioecious animals. In many aquatic pulmonates copulation is polarized, with one individual acting as male and the other as female. It is especially clear here that if being the male resulted in more offspring per unit cost, maleness would be favored by selection until the diminishing supply of fertilizable ova forced the more persistently male individuals to greater and more costly effort. There seems to be no factual justification for

121

Maynard Smith's view that internal fertilization in outcrossed hermaphrodites avoids the cost of meiosis.

I would also take issue with some aspects of Ghiselin's (1969) reasoning. He proposed three general groups of theories to explain the incidence and kinds of hermaphroditism in the animal kingdom. One, the Low-Density Model, is based on the advantage of being able to reproduce at population densities so low that conspecific individuals may be hard to find, and those of a specified sex twice as hard to find. Selection favors simultaneous, self-fertile hermaphroditism because it is uniparental. The second is the Size-Advantage Model for sequential hermaphrodites, based on the likelihood that optimal size for male and female functions may be different. The third, Gene Dispersal Model, is based on the advantage of avoiding inbreeding or the unfavorable genetic consequences of small population size. A remnant or founder group of, say, six hermaphrodites will have an effective population size (at a crude theoretical level) of six. A group of six dioecious individuals can have no greater an effective population size and will probably have a smaller one, as low as zero when all six are of the same sex.

In two respects I believe that Ghiselin's valuable presentation fails: first of all, its reasoning on adaptation relies too heavily on assumptions of optimization that are insufficiently explicit in their Darwinian basis; and secondly, the level of taxonomic attention (major groups of the animal kingdom) is not likely to produce critical evidence for testing the theories. A good example of the difficulty with general optimization models is Ghiselin's Sampling Error version of the Gene Dispersal Model. He points out that, compared to a dioecious alternative, an individual in a restricted population of hermaphrodites is favored because it and its descendants will be less affected by inbreeding depression. This advantage, of being a member of a group of hermaphrodites, is irrelevant to selection. The relevant advantage would be one that would belong, in a group with a tendency towards hermaphroditism, to individuals with a greater than average degree of hermaphroditism. There is no indication that this would be true. Ghiselin specifically warns against comparing explanations that relate to different levels of organization and at one point rules out the applicability of group selection. To me his Gene Dispersal Model looks group selectionist.

Significant variation in reproductive patterns often occurs only at high taxonomic levels in the animal kingdom, and this reduces the usefulness of the information as it relates to the adaptive significance of hermaphroditism. The whole class of oligochaetes are simultaneous outcrossed hermaphrodites. It is unlikely that the forces that led originally to their hermaphroditism are still at work throughout the group. Rather it must be that they lack preadaptations for changing to some other mode of reproduction. Critical evidence on such theories as Ghiselin proposed are more likely to come from studies of groups in which reproductive patterns vary widely at the level of species and genera.

CHAPTER ELEVEN

Why Are Males Masculine and Females Feminine and, Occasionally, Vice-Versa?

The essential difference between the sexes is that females produce large immobile gametes and males produce small mobile ones. In nonreproductive aspects of life, a male and a female of the same population usually have similar ecological relations: roughly the same habitats, diets, diseases, and so on. The ultimate goal for both is maximal genetic representation in the same population. Yet their immediate goals in reproduction may seem strikingly different. Males of more familiar higher animals take less of an interest in the young. In courtship they take a more active role, are less discriminating in choice of mates, more inclined toward promiscuity and polygamy, and more contentious among themselves.

Of all these aspects of masculinity, only the last item has gotten serious attention, and even the best works on sexual rivalry, from Darwin (1871) to O'Donald (1972) mainly take the general roles of the sexes as something given, and direct attention to tactical details of sexual rivalry. The one comprehensive attempt to provide a general explanation for the forms and distribution of masculinity and femininity (Wynne-Edwards, 1962) I regard as totally erroneous (Williams, 1966A).

The masculine-feminine contrast is a prima facie difficulty for evolutionary theory. Why, if each individual is maximizing its own genetic survival, should a female be less anxious to have her eggs fertilized than a male is to fertilize them, and why should the young be of greater interest to one than to the other? This chapter attempts to show that different opti-

124

mal strategies result from the gamete size difference, such that X
masculine characters are usually optimal for males, and feminine characters for females. It deals in turn with three phases
of sexual reproduction: gamete production, achievement of
fertilization, care of offspring.

GAMETE PRODUCTION

Over part of the possible range of fecundity it must be true
that the more gametes are produced, the more descendants will
result, but the relationship must sooner or later break down.
If no other factor intervenes, depletion of parental resources
will. There must be an optimum level of expenditure on reproductive functions, dependent on the relative values of immediate and long-range reproductive interests, and on an optimum
apportionment of this expenditure among different aspects of
reproduction. Optimization of total effort for different species,
but not for different sexes, was discussed by Gadgil and Bossert
(1970) and by Williams (1966B).

Reproduction of sessile marine animals often requires little
beyond production and release of gametes, and here we expect
a minimum of contrasting masculinity and femininity. Fertilization is achieved by nearly synchronous release of eggs and
sperm, or by release of sperm and their transport to exposed
but attached eggs. There is no reason to doubt that 10% more
eggs would mean 10% more zygotes. Similarly the production
of 10% more sperm will raise by 10% the probability that
a given egg will be fertilized by one's own sperm and not by
someone else's or not at all. This is true except in the probably
infrequent situation of most of the nearby eggs being already
fertilized by one's own sperm. Unless there is brooding or
viviparity, males and females of a sessile marine invertebrate
should spend equally on gamete production. I know of no data
on this point, but Kalmus and Smith (1960) state that sessile
marine invertebrates have similar size ovaries and testes, and

this is certainly the impression one gets from gross dissections of such animals as sea urchins. Data on seasonal size changes in oyster gonads are normally given without specifying the sex of the individuals studied (Galtsoff, 1964; Sastry and Blake, 1971).

Mobile animals achieve fertilization by close coupling and consequent reduction of sperm wastage. A small amount of material devoted to sperm production may be adequate to assure fertilization of all the eggs of as many females as are likely to be available. For the female it may remain true that output of offspring is linearly related to egg number. So in mobile animals the mass of eggs should greatly exceed that of sperm, except as this relation is complicated by such factors as viviparity. Comparison of gonad sizes in oviparous fishes confirms this expectation, with testes generally between a half and a tenth the size of ovaries in specimens of comparable size (Gunston, 1968; Heally, 1971; Hoyt, 1971; Kaya and Hasler, 1972; Krumholtz, 1958, 1959; Mathur, 1962; Norden, 1967; Otsu and Hansen, 1962). The further reduction in testis size in fishes with internal fertilization confirms efficiency of sperm use as the essential factor. Moser (1967) recorded a 10- to 20-fold difference in gonad size in males and females of the internally fertilized *Sebastes*. The ideas proposed here indicate no reason why a large male should have a larger testis than a small male. Gunston (1968) found that relative testis size decreased with increasing body size of salmon.

In mammals the cost to the female comes almost entirely after fertilization, and this may be true to a lesser extent in birds. Reproductive efficiency demands that fecundity be appropriate to resources likely to be available later to the young. In bird populations studied by Lack (1954, 1968) he was able to show that increased egg fecundity beyond an optimal level results in a lower, not a higher number of fledglings produced, or at least in fledglings of reduced weight and survivorship. Comparable indications were found for mammals. These conclusions have been both challenged (Wynne-Edwards, 1962) and supported (Ricklefs, 1968) by many others. Female fecundity in all animals in which cost mainly follows fertilization will be adaptive in relation to the rest of the reproductive program and may be well below that which is physiologically achievable. In these organisms we find, as expected, small testes and in mammals an essentially testis-size ovary.

ACHIEVEMENT OF FERTILIZATION

Differences in expected male and female strategies would again be minimal for a sessile marine invertebrate. Both sexes can reproduce merely by releasing gametes at the same time as one or more individuals of opposite sex. Synchronization can be achieved by timing by readily perceived physical factors such as light or temperature, whenever these are predictive of optimum timing of reproduction. More precise synchrony is possible if one individual can signal the start of its spawning to others of the opposite sex. The male would be more effective in giving such a signal. Sperm are readily transported by currents, along with possible chemical stimuli present in the semen. A perceiving female can then release her eggs with a high probability of fertilization. If she spawned first, the eggs could sink or float and not be readily detected by males. A chemical stimulus released with such eggs would diffuse in a way different from the scattering of eggs and be poorly predictive of their presence. It is understandable that a general synchronization by physical cues, with prior spawning by males as an immediate stimulus to females, is the general rule among sessile marine invertebrates (Thorson, 1950). Taking the initiative in spawning is the sole expression of masculinity in these animals.

Mobile animals, with copulation and efficient use of sperm, show more pronounced masculine-feminine differentiation. Fertilization of nearly all eggs produced by one or even several females requires a negligible expenditure on gametes by the male. His reproductive success is mainly a function of the number of females available to him, and selection will favor any character that increases the number of successful matings. Two kinds of such characters are expected, those that reduce the influence of competing males, and those that stimulate sexual cooperation by females. These are the much discussed characters developed by sexual selection (Faugeres, Petit, and Thibout, 1971; O'Donald, 1972).

Mate quality would seldom be as important to a male as their number. Where female fertility is strongly size dependent, as in fishes, a male should prefer larger to smaller females. Human female fertility is strongly age dependent, and the value of copulation for the male would be described by the age-distribution of the female fertility. If completely promiscuous behavior were the rule, the age distribution of female attractiveness should correspond exactly to that of female fertility, with a peak in the early twenties. If strict lifetime monogamy prevails, the male strategy should be to pick a female with maximum reproductive value (total expected future reproduction), so as to realize maximum return on a lifetime commitment. He should be maximally attracted to an adolescent girl. If prolonged pair bonding with some possibility of divorce and adultery is the rule, the distribution of female attractiveness should lie between those of fertility and reproductive value. This is undoubtedly the realistic average marital system, and should produce a peak of female attractiveness in perhaps the late teens.

There seem to be no data on this point, but my impression is that a strong male preference for women who are sexually mature but youthful is universal in all cultures, perhaps more so than any other standard of beauty. Male age should be a less important factor in the woman's choice of mate. Individual variation in sexual preference indicates the great plasticity of human behavior, but in no way impugns the evolutionary significance of the norms. Emlen (1966) has written a stimulating discussion of the natural selection of human sexual and social behavior.

If one mating can provide enough sperm to fertilize all her eggs, the only advantage to multiple matings for a female would lie in benefits from increased offspring diversity. This could only be a minor factor, because even in a completely monogamous system, within-progeny variance is half that of the whole population. Mating with a very few males would give very nearly the maximum attainable diversity from sexual reproduction. It would appear that most female fishes do not release all eggs during a single copulation, but spawn several times during the breeding season, often with different males (Breder and Rosen, 1966).

Mate quality rather than number must be the dominant consideration in female strategy. For her the important consideration is to choose the optimum moment and circumstances for converting her eggs to zygotes. In many species it would seem that matings with other species would frequently occur if it were not for female choice. Bateman (1949) noted that Drosophila species with effective ethological isolating mechanisms may hybridize readily if anesthetized females of one species are presented to males of another. Males of various poeciliid fishes readily court females of other species, and conspecific matings are the rule only because females reject males of all but their own species. Only after a period of learning do males show preference towards females of their own species (Liley, 1966).

The principle that one sex tries to achieve fertilization to the maximum degree and the other tries to optimize it in a qualitative sense is a key to understanding many aspects of courtship and sexual behavior. Trivers (1972) was the first to see the full implications of this principle. He pointed out that the seemingly cooperative aspects of family life may be misleading, if they suggest that the male and female have identical interests in the outcome. At various points in the reproductive program, it may be especially advantageous for one member to cheat in certain ways (adultery, desertion, etc.) and at times it may be especially disadvantageous to have one's mate indulge in such activities. Trivers interprets much of courtship and related behavior in birds as attempts to preserve one's own options and restrict those of the mate. Courtship of course has other functions. Its role in preventing interspecific matings is well documented.

Another difference in sexual strategy is seen in the canalization of reproductive functions. Female functions must be precisely coordinated because one malfunction may produce a great waste of resources. In males, even gross abnormalities such as homosexuality, or courtship behavior at inappropriate seasons, will result in trivial losses and not be strongly selected against. Normally the only serious loss would be a missed opportunity for a productive mating. A review of evidence that male reproductive functions are poorly canalized is provided by Williams (1966A). Additional data on such points as the greater frequency of male homosexuality, nonfunctional juvenile sex behavior, and substandard parental behavior are provided by Andrew (1966), Beach (1964), Bourliere (1964), and Heard (1972).

I once concurred (Williams, 1966A) in the common view recently advocated by Orians (1969) that female discrimination favors the genetically fittest male, as this fitness is expressed through his phenotype. Supposedly she can thereby optimize the genotypes of her young. I now feel that this would require an unrealistically high heritability of fitness. The important female adaptation in relation to courtship is an ability to predict future resources for her offspring from the appearance and circumstances of a courting male.

Even in monogamous species, in which opportunity for fertilizing eggs of more than one female is usually absent, the expected masculine-feminine contrasts should appear. Monogamous pairing through a complete reproductive program may be merely the one stabile combination among several false starts. Even if both parents contributed equally to the rearing of the young, desertion by the father right after fertilization would have a more adverse effect on the mother's interests than his own. He can quickly replace wasted gametes and be ready for another mate. She can not so readily replace a mass of yolky eggs or find a substitute father for an expected litter. Her optimum behavior prior to copulation should show a much greater degree of caution than the male's.

The contrast should be more strongly developed in polygynous species, in which males usually contribute no support for the offspring. The greater masculinity should consist mainly of greater development of characters used in competition with other males. Discrimination by the female should be more directly related to habitat resources. Male success would be mainly dependent on how many females he can keep inaccessible to other males. This will largely depend on the size of his territory and its attractiveness to females.

The importance of territory suitability for attracting females is indicated by those birds that show greater female fidelity to a territory than to a mate (Varner, 1964). Trivers (1972) reviews other instances of female choice that relate more to the locale than to the attendant male. Female dragonflies choose the best oviposition sites, and mate with whatever male happens to be available at the site. Male rivalry shows mainly in competition for territories that include good oviposition sites, and courtship consists of advertizing the site.

The literature is full of descriptions of male-female differences in reproductive behavior in accordance with the ideas expressed, including the expected differences between monogamous and polygynous species: males are first on the breeding grounds; polygyny is common, polyandry rare; nonfunctional courtship and mating, such as homosexuality, is more common in males; it is the males that have elaborate courtship displays, etc. Interpretations vary from thoroughly erroneous to the enlightened treatments by Downhower and Armitage (1971), Orians (1969), Selander (1965), and Trivers (1972).

Downhower and Armitage provide a well-analyzed study of contrasting male-female strategies. They found that a female ground squirrel would be most successful as a mother when she had a male and his territory all to herself. Members of harems were less successful, and mean success in harems of two was greater than that in three, the largest observed size. As expected, females show a sexual strategy designed to minimize harem size. They distract males courting other females and behave aggressively towards rivals. For the male, as expected, the more wives he has the more offspring he produces, up to a point. Calculations indicated that a harem size of about 2.5 would be optimal for a male. The outcome was a modal harem size of two, a somewhat male-favored compromise.

REVERSED STRATEGIES IN RELATION TO FERTILIZATION

An obvious test of this explanation of masculine-feminine contrast is provided by those animals in which the male advantage in multiple copulation and female advantage in optimum timing and mate selection would be partly reversed. In the seahorse-pipefish family Syngnathidae, copulation results in transfer of ova to the male, who becomes pregnant with embryos partly dependent on a placental attachment. Copulation commits a male to a prolonged burden, and there must be a limitation on the number of females whose eggs can be accommodated by one male. The limit may often be one. In a seahorse (Strawn, 1958) and a pipefish (Takai and Mizokami, 1959) the ovaries of a female have on the average about the same number of eggs as the number of embryos normally found in a male brood pouch. The small size of the young, and the fact that some yolk remains in the yolk sac at birth, suggest that much of the nourishment of the embryo is provided by the mother. Only a detailed study of reproductive physiology can settle the matter of relative cost to male and female.

Most of the extant information comes from casual observation, often by amateur aquarists, but it appears that the theoretical expectations are realized. There are numerous reports of females being more brightly colored and aggressive than the males. This picture emerged clearly for the three species carefully studied by Fiedler (1955). General reviews of what is known of reproduction in this family (Breder and Rosen, 1966; Herald, 1961) support the expectation that at least partial reversal of sex strategy should be found. Undoubtedly a detailed study of reproduction in these fishes would show that both physiological burdens and behavior patterns are highly varied. I predict that the behavioral masculinity and femininity will closely parallel variation in parental burdens of male and female.

Other forms of egg holding by males are found in Pycnogonids (Hedgpeth, 1963), some anurans, and a few insects. The burden for the male midwife toad is merely mechanical, but there must be some limit to the number he can carry. We would expect a reduction in courtship activity as the burden increases, and perhaps an increased courtship by the female, but apparently little is known about reproductive behavior in this species. Dahne (1914) reported that the female had a mating call and that it was louder than the male's. This was accepted by Noble (1931) but ignored by later writers. Nothing is known of sexual behavior or seriousness of the male burden in those water bugs in which the males carry eggs. This phenomenon in dendrobatid frogs is a brief transport of hatching eggs or tadpoles from the nest to water and probably costs very little (Stebbins and Hendrickson, 1959; Savage, 1968).

In some fishes the fertilized eggs are carried in the mouth of the mother and in others that of the father. Oral incubation by males, although a milder form of pregnancy than that of the syngnathids, would be expected to produce some reversal of sex strategies. It is suggestive that in species with female brooding, males show bright nuptial colors and females remain dull. With male brooding or cooperative nesting, the sexes are similarly colored (Baerends and Baerends, 1950; Myrberg, 1965).

A rare but taxonomically widespread phenomenon among birds is the male's assumption of the entire burden of incubation. These reverse-role birds are conveniently reviewed by Kendeigh (1952) and, along with some doubtfully comparable crustacean developments, by Wynne-Edwards (1962). In a few examples there is a thorough reversal of the sex roles in reproduction: females polyandrous, more brightly colored, more active in courtship. In others the reversals are partial. For instance, brushland tinamou males and females have similar plumage, polyandry is rare, and males are, at least initially, more active in courtship (Lancaster, 1964).

Even partial reversals of masculinity and femininity occur only where males assume burdens usually taken by females. This is strong evidence for the view that in most species masculine characters are individually optimum for males and feminine ones for females. No benefit-to-the-species interpretation of reproductive behavior need be invoked. This is also a choice illustration of the logic of the comparative method. If there were no such animals as seahorses or tinamous, the explanation offered would be compatible with the evidence but not forcefully supported. It is the exceptions to the rule of masculine males and feminine females that prove the theory that explains both rule and exceptions.

CARE OF THE YOUNG

Kin selection may promote the evolution of parental care of the young, if the necessary preadaptations are present. Male territoriality and external fertilization preadapt a species to the evolution of parental care by the male. Internal fertilization without territoriality preadapts the species to parental care by the female. Continuous association of the parents after fertilization permits the evolution of parental care by both sexes.

Most bony fishes have external fertilization, and male territoriality is common. The function of territoriality is a currently debated issue, but at least it is clear that larger territories in preferred habitats favor reproductive success over those that are smaller or in marginal habitats. Typically in territorial fishes the eggs are laid and remain in the territory. The female is driven away, but the male remains associated with his offspring, and defense of territory becomes incidentally a defense of young. Tendencies to defend against predators other than rival males, to avoid harmful disturbances to eggs already laid, to refrain from eating eggs or young, to keep them free from parasites, etc., are readily evolved by kin selection. This explains the high frequency of paternal tending of externally fertilized eggs in fishes, and the rarity of such behavior by females.

Emlen (1973:148) suggested another explanation. He reasoned that indeterminate growth and dependence of female fertility on size will place a special premium on the female's use of resources for further growth. She can presumably grow more rapidly without the burden of nest guarding and is selected to leave the eggs entirely in the care of the male. Unfortunately for this argument, there may be an even greater advantage for the male in growing larger. Size can be important in competition for territories and in courtship. The fact that males are often larger than females in nest tending species indicates that large size is more important for the male.

A male fish guarding a nest of eggs has a task largely unrelated to their number. It must be nearly as demanding to defend a hundred as a million. This is true wherever eggs require neither incubating nor carrying, and the young require no feeding. A wasp guarding the entrance to her burrow has the same task no matter how many eggs lie behind her. This has no doubt been a factor in the evolution of cooperative nesting in this group (Hamilton, 1964A, 1964B). When the young require feeding, this advantage in cooperation no longer applies.

The ease with which a male fish can tend large numbers of eggs in his territory, if he tends any at all, explains the lack of parallelism between this kind of parental care and that shown by syngnathid fishes and male-brooding birds. The nesting fish retains interest in receptive females, and courtship may alternate with nest tending for a while. Only when the season of occurrence of spawning females comes to an end does the male become exclusively the father and no longer the suitor. In such species the adaptive value of spawning only under optimum circumstances remains high for females, and the value of multiple mating remains high for males. Paternal nest tending in fishes leads to little reduction of masculine-feminine contrasts.

Cichlids are among the few fishes in which both sexes are territorial, and in these we often find a sharing of family responsibility. One such responsibility is egg cleaning in the mouths of the parents. From this momentary egg carrying, oral incubation has evolved independently several times. When it is the female that carries the eggs she remains thoroughly feminine and her mate masculine. When the male carries them there is some reversal of sex roles in courtship. Aronson (1948) maintained that females of the male-brooding *Tilapia malanotheron* are more active in courtship. Barlow and Green (1970) deny this, and say that the smaller fish, whatever its sex, is more active. This makes tactical sense. A larger fish will presumably invest a larger absolute measure of resources in raising the brood, but the genetic payoff is divided equally between the parents. The smaller fish should be more desirous of the cooperation of the larger than vice versa. It would appear that male masculinity is at least reduced in this species, if courtship roles are decided more by size than by sex. There is sexual dichromatism, but both sexes have nuptial markings and it is difficult to decide which is more extreme.

Internal fertilization as it occurs in most land animals means that the female remains associated with the young longer than the male, a preadaptation for the evolution of maternal behavior. When the male is territorial and the female remains in his territory after fertilization, as she must if she is a brooding bird and may if she is a mammal, care of the young by both parents may evolve. This is the pattern in many birds and the Carnivora among mammals.

I regard the sharing of guard duty, as it occurs in the cichlids and some catfishes (Breder, 1935) as a serious evolutionary problem. Birds bring needed food to their young, and two should bring about twice as much as one. Selection for parental behavior could be nearly the same for the two sexes. As Emlen (1973:148) clearly recognized, mere guarding could be done almost as well by one as by two. The removal of one of a pair of fishes nesting in an aquarium causes no adjustments by the one remaining. It merely continues to tend the eggs and may be entirely successful in preventing losses to other fishes or to snails or to the fungal infections that may decimate untended eggs. Also it is unlikely that loss of the brood would be equally harmful for both parents. The female would have the more prolonged and expensive job of replacing her contribution. So the sharing of guard duty should be unstable. Males should be selected to leave the eggs to the female and seek another mate, as discussed for birds by Trivers (1972).

Even in some mouth-breeding cichlids the eggs are shared by the parents (Afpfelbach, 1966; Myrberg, 1965), and the larger parent takes a larger share of eggs. I would expect the evolution of male tendencies to desert the eggs and court additional females, but the egg sharing and similar sharing of re-

sponsibilities found in other groups must be a stable evolutionary equilibrium. At fertilization the males have invested little
more than gametes, so that the factor of a cumulative investment to be protected, as envisioned in Trivers' models, seems
not to apply. The continued availability of spawning females
is highly probable in these tropical fishes with extended breeding seasons. It may be that the explanation will emerge from
consideration of not just one but a sequence of broods. These
fishes have prolonged pair bonding in captivity. It may be
that a suitable territory held jointly with a cooperative female
is a long-term resource so hard to win that, once secured, it
should not be jeopardized for a possible immediate gain from
courting another female.

SEX AND MORTALITY

I believe that Trivers (1972) gives the correct explanation
for why males have greater mortality rates than females. He
easily discredits the myth that whichever sex is heterogametic
has higher mortality rates from vulnerability to deleterious recessives on the single X-chromosome, and argues that the sex
with greater variance in fitness should have greater mortality.
In many species a typical adult female will enjoy something
like the mean reproductive success. A male, especially in
polygynous species, may not reproduce at all. Perhaps only
the fittest 25% of the males will reproduce, and the top 1%
may enjoy many times the mean reproductive success. At every
moment in its game of life the masculine sex is playing for
higher stakes. Its possible winnings, either in immediate reproduction or in an ultimate empire of wives and kin, are greater.
So are its possibilities for immediate bankruptcy (death) or
permanent insolvency from involuntary but unavoidable celibacy. Trivers discusses a number of observations of the greater
vulnerability of males. Greater male susceptibility to psychosomatic damage is documented by Murdoch (1966). Giesel

(1972) reviewed evidence that male insects are less well canalized physiologically and less tolerant of environmental stress.

[Greater variance in male fitness not only affects optimization of reproductive behavior, but all tributary aspects of adaptive organization] A male's developmental program must gamble against odds in an effort to attain the upper tail of the fitness distribution. A female's need merely canalize against malfunctions [Female mortality will be found to exceed male, not in species with female heterogamety, but in those with female masculinity.]

CHAPTER TWELVE

Sex as a Factor in Organic Evolution

If genetic recombination can be an immediate reproductive adaptation (Chapters 2–5), appeals to long-term benefits of evolutionary potential are inappropriate in an explanation for the origin and persistence of sex in a population. On the other hand, it is unlikely that sex could be irrelevant to the evolutionary process. Its role in evolution is a problem of basic importance, formally separable from that of sex as a character shaped by selection. It is difficult, on the basis of brief statements in textbooks, to decide exactly what is the generally accepted view. Benefits of sexual reproduction are often said to lie in the production of a diverse array of genotypes for selection to act upon. An example is Mayr's (1963) statement:

> Through recombination a population can generate ample genotypic variability for many generations without any genetic input (by mutation of gene flow) whatever.

A closely similar statement is made by Dobzhansky (Tax and Callender, 1960:115).

In my opinion this popular view is the one most likely to be correct—ample genotypic variety may be a safeguard against extinction—but my reason is the opposite of that usually assumed. The reason usually given is that recombination increases responsiveness to selection and the rate of adaptive change. This chapter proposes that sexual reproduction usually opposes the effect of selection, and the final chapter proposes that this retardation of adaptive evolution may produce a long-term group advantage. Neither view of the role of sex in evolution is supported by any of the serious theoretical work that has been done, and Kimura and Ohta (1971) argue that such views as Mayr's and Dobzhansky's are contrary to all important thought since Fisher and Muller.

In this mainstream of theory the conclusion usually supported, and the only one seriously considered, is that recombination increases the rate at which favorable mutations can be incorporated in an evolving population. The benefits of sexual reproduction are said to relate to long-term advances that arise from steady replacement of an original germ plasm by new mutations that are favorable when they arise and remain so as other favorable mutations arise at other loci. As Kimura and Ohta expressed it: "Sexual reproduction has played a very important role in speeding up evolution in the past, helping to produce man before the sun in our solar system burns out."

THE MULLERIAN THEORY

Recent thought on the evolutionary effects of recombination is in a bizarre state of contention after forty years of uncritical complacency. The complacent period began with R. A. Fisher (1930) and especially H. J. Muller (1932), whose classic essay begins optimistically with the claim

> . . . that genetics has finally solved the age-old problem of the reason for the existence (i.e., the function) of sexuality and sex, and that only geneticists can properly answer the question "is sex necessary."

and then goes on to develop the view now generally accepted by population geneticists. It assumes that sex can scarcely be advantageous to the reproducing individual and must be produced and maintained by group selection. Even if there were no general objections to the concept of adaptation through group selection, there would be an adequate one here. The existence of populations in which asexual and sexual reproduction occur in evolutionary stability demonstrates that sex gives some advantage that balances recombinational load and the cost of meiosis.

Muller and Fisher's theory proposes that sex is favored by group selection because it should increase the speed with which a population can incorporate favorable mutations. With asexual reproduction, a mutation that arises in one individual can never be present in members of another clone. With sexual reproduction, two individuals can have descendants in common, and genes from different sources can come together in common descendants. Recombination produces genotypes that would not otherwise arise, and putting two favorable mutations together supposedly increases the selection of both.

This theory was modernized by Crow and Kimura (1965) who summarized the Mullerian view in a graphic model (Figure 14). Their publication sparked the beginning of a renewed interest in sex as an evolutionary factor. At about this time the Wynne-Edwards theory of group selection was being warmly debated, and Maynard Smith (1964) was among those who rejected the theory in its application to reproductive physiology and social behavior. It was his thinking on group selection in relation to reproduction and population regulation that led him to consider it as a possible factor in the origin of sexual reproduction (Maynard Smith, 1971b).

Crow and Kimura found that sex is of no importance in accelerating evolution in populations of less than a thousand, but that it can enormously accelerate adaptive change in really large populations. They confirmed Muller's reasoning that a favorable mutation could only rarely be established in a small population. By contrast, even rare mutations can occur at finite frequency in enormous populations. Sexual reproduction could bring favorable mutations together so that they could enhance, rather than compete with each other.

Although they seemed to confirm the Mullerian view, Crow and Kimura really made that view a bit suspect. Theories of group selection rely heavily on populations being numerous

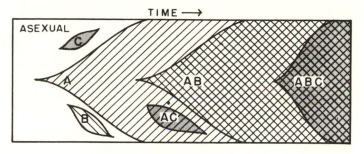

LARGE POPULATION

SMALL POPULATION

FIGURE 14. Model of incorporation of favorable mutations in asexual and sexual populations, from Crow and Kimura (1965). A, B, and C are favorable mutations and all three independently replace ancestral alleles in the large sexual population. In the large asexual population, clones with mutations B and C prevail over the ancestral genotype for a while, but die out in competition with clone A. Only when B and C arise again in a member of clone A can they be incorporated. In the small populations favorable mutations are so rare that each completely replaces competing alleles before the next is likely to arise, and this is true regardless of whether reproduction is asexual or sexual.

143

and small, rather than large. The requirement that population size be well over a thousand removes the theory's applicability to some interesting organisms. This requirement has been drastically revised upwards by later work and is discussed further below. The theory also implies assumptions on gene interaction in the determination of fitness that would not be universally accepted as the norm for favorable mutations.

These are serious problems, and so is that pointed out in a criticism by Maynard Smith (1968A; see also reply by Crow and Kimura, 1969). He argued that Crow and Kimura are correct only if the favorable mutations are unique events. With a finite rate of origin for each mutation, and with the assumption that mutuations at different loci occur independently, it is a simple matter to calculate a frequency for each genotype. The frequencies will be exactly the same, whether reproduction is asexual, sexual, or a mixture of the two. Sex would have an effect only in a population in which genotype frequencies did not conform to the "independence relation" (linkage equilibrium) expected to arise by mutation pressure This could happen from the mixing of separate populations. So Maynard Smith concluded that sex is important in facilitating adaptation to new habitats invaded by colonists from divergent sources.

More recently Maynard Smith (1971A) reinvestigated the problem in greater detail. He confirmed his earlier view of the importance of sexual reproduction among colonizers, but also found support for a modified form of the Mullerian theory. He concluded that sex would be of negligible long-range importance with population sizes less than about ten times the reciprocal of rates of favorable mutation. Thus if a typical favorable mutation arises in one in a million gametes, the population would have to be larger than ten million for sex to accelerate evolution. In an infinite population, sex would accelerate evolution by a factor equal to the number of loci at which favorable mutations can occur. R. A. Fisher had reached this last conclusion without recording his calculations, but did not appreciate its inapplicability to finite populations.

Kimura and Ohta (1971) also introduced a more stochastic model as a refinement of the one by Crow and Kimura. It concurred in the conclusion that populations would have to be in the millions for recombination to be important in increasing the spread of favorable mutations. Kimura and Ohta also point out that recombination is disadvantageous when fitness depends on heterosis or epistasis.

Bodmer (1970, 1972) considered early stages of incorporation of favorable mutations and calculated that sexual reproduction may acclerate the formation of recombinants by a factor of two or more at each locus. This advantage would be additive among loci and greater for small populations than for large. Bodmer proposed this as the reason for sexual reproduction being more common in eukaryote than in prokaryote populations. It is only in the eukaryotes that populations would ordinarily be so small that there would be a significant benefit from recombination. Bodmer's conclusion on the effect of population size is the exact opposite of everyone else's. It is my understanding that Joseph Felsenstein is preparing a treatment of this matter which points out an error in Bodmer's reasoning.

Bodmer also recognized the advantages of asexual reproduction as a way of preserving and multiplying highly fit combinations. He suggested that the optimum evolutionary potential may consist of some combination of asexual and sexual reproduction. The same suggestion is made from time to time by botanists, and theoretical reasons why the combination should be advantageous were discussed by Wright (1956)

Eshel and Feldman (1970), alone among recent contributors, have proposed that evolution may normally be retarded in exclusively sexual, as opposed to exclusively asexual populations. They find the production of new gene combinations by meiosis and fertilization to be in some circumstances a less

145

potent force than the breaking up of adaptively proven combinations by the same process. Their work is discussed in relation to my own views below.

Unless Bodmer is correct, and everyone else wrong, the Mullerian theory is obsolete as a broadly applicable principle. Specifically, the requirement that population sizes be in the millions or more for the proposed evolutionary benefit to be of any consequence removes the benefit from a large fraction of the Earth's biota, including its most rapidly evolving members. It is of no relevance, for example, to human evolution at least since the Devonian. We may have to go back to the placoderm or even protochordate stage to find populations big enough among man's ancestors for sexuality to accelerate evolution. Any of the commoner members of the marine plankton are able to incorporate favorable mutations thousands of times as rapidly as any of man's known ancestors, yet many of them today are scarcely different from their ancestors in the English chalk. Despite the reasons why man's evolution ought to be slow—small populations, low mutation rate, long generation length—he evolved with unusual rapidity.

ADEQUACY OF EVOLUTIONARY RATES

Evolutionists at the time of Lamarck and Darwin, who underestimated the age of the Earth by orders of magnitude, were understandably anxious about the adequacy of evolutionary rates to account for the observed diversity of organisms. Similar motivations still prevail in modern biology and form an important aspect of criticism of evolutionary theory (mostly by nonbiologists) in a recent symposium (Moorhead and Kaplan, 1967). Another example is Waddington's (1957, 1959) proposal on genetic assimilation as a factor in evolution. He proposes that without genetic assimilation, evolution would not go fast enough. This concern with evolutionary rate is also the central consideration in all recent works on sex as an evolutionary factor. They are all primarily concerned with the question: What are the relative rates of selective gene substitution in asexual and sexual populations? They are a recent example of the often discredited view that rates or directions of evolution are determined by internal machinery, rather than by environmental demands.

For the important evidence we need look no further than Darwin's work on domesticated animals and plants. Here we see that organisms suddenly removed from nature's to man's world may adapt to the new ecological niche with rates of evolution perhaps a thousand times that which had been occurring. This rapid evolution under domestication characterizes plants and animals, newly evolved and ancient taxa, and organisms with widely different life cycles. The same sorts of observations have more recently been made on organisms of interest to hobbyists or response-to-selection experimenters. The obvious conclusion is that organisms can evolve as fast as nature is ever likely to require; slow evolution must mean slowly changing requirements. The "living fossil" *Limulus* has a normal level of genetic variability and presumably would respond rapidly to artificial selection (Selander, et al., 1970). If it were readily cultured, I suspect that man could accomplish more change in a few decades than nature has demanded in a hundred million years. Extinction occurs not because an organism loses its adaptation to an ecological niche, but because its niche becomes untenable, as discussed in Chapter 13.

Much of the textbook level of theory on evolution is still devoted to showing that the postulated factors are capable of producing sufficiently rapid change. They ought to be concerned with explaining how populations can so effectively resist change over such amazingly long periods. Genetic processes in the evolutionary context are presented as the machinery of change, rather than of stability. Mutations are considered the raw material of evolution rather than functional mistakes. Recessiveness in diploid populations allows them to accumulate genetic diversity so that they can respond sufficiently rapidly to later environmental change. And so on.

Serious recent works on diploidy and dominance (Crow and Kimura, 1965; Maynard Smith, 1971B) adequately show the fallacy of the variation-storage proposition. Diploidy is favored because of immediate benefits from heterosis and the suppression of effects from deleterious elements in one or the other genome. Dominance arises from selection for canalization, and completely recessive genes are expected to be consistently deleterious. Also the initial stages of selective increase of recessive genes are extremely time consuming. Recessives in low frequency are an ineffective raw material for rapid evolution.

That mutability is maintained by the advantage of having evolutionary raw material is still seriously considered by reputable theorists (Kimura, 1967; Levins, 1967). The idea is that mutation rates represent a compromise between the need for short-term stability and that for long-term evolutionary plasticity. I have argued (Williams, 1966A) that observed mutation rates should all be regarded as approximations to zero, with variation reflecting intensity of selection for minimizing mutation. Kimura rules out this possibility on the basis of nature not having produced a uniformly low mutation rate. It can also be observed that not all cryptic animals are uniformly invisible, nor all cursive ones uniformly fleet. Does this imply an intermediate optimum of visibility or attainable speed? Field biologists recognize many taxonomically variable characters as having optima at zero or infinity, why not the geneticist?

Given sufficient time and selection pressure, I assume that mutation rate could be reduced to any specified value. In nature the stability of a gene and its influence on the stability of other genes are merely two among many factors in its selection. Mutation rate should vary predictably, according to the likelihood and extent of fitness reduction by mutation. It should be lower per life cycle in haploid than in diploid organisms, and lower in diploids than in polyploids. It should be lower per unit time in long-lived than in short-lived organisms. It should be lower at loci at which prevalent mutations are lethal than at those at which nearly neutral mutations are the rule. It should be lower in historically normal ecological and genetic environments than under unusual conditions. I reviewed evidence in support of these expectations in an earlier work (Williams, 1966A), and Leigh (1970) has advanced additional arguments in favor of the zero optimum. Levins (1967) found mutation rates consistently lower than those predicted by his optimization models.

Generation length should certainly have a strong influence on potential rate of evolutionary change. Drosophila would give more rapid results in a response-to-selection experiment than man, but in the last few million years man has evolved more rapidly than Drosophila. Simpson (1950) pointed out this lack of a relation between generation length and rate of evolution and that mammals evolved faster than forams, despite the much longer generations of mammals. The evolutionary significance of sexual reproduction must reside in effects other than those on potential rate of evolutionary advance.

SHORT-TERM EVOLUTION IN ASEXUAL AND SEXUAL POPULATIONS

The last section indicated that potential rates of gene substitution are always many times greater than those normally demanded by the environment. So the question as to whether asexual or sexual populations can more rapidly accomplish gene substitution is perhaps logical, but not biologically very important. It may be more appropriate to inquire into another kind of evolutionary change, that of short-term shifts in Hardy-Weinberg deviations and linkage disequilibria. These

149

are matters of recombination on which sexual reproduction has an immediate and powerful effect.

Suppose the question is posed: How long does it take for the best available genotype to become the norm in asexual and sexual populations after an environmental change? The obvious answer, valid as a first approximation, is this: In the asexual population the best available genotype becomes the norm immediately; in a sexual population it never does. Of course the best available genotype in the asexual population may not be as good as some that could be produced by recombination, but the occasional production of extremely fit genotypes in the sexual population will have no permanent significance, as long as fitness depends at all on heterosis and epistasis. Given any plausible level of heterosis and complex interactions among loci, almost all of a sexual population will have suboptimal genotypes, no matter how long selection continues. All that this means is what everybody knows, that sexual reproduction generates recombinational load, but I suggest that this mundane fact may be the primary significance of sexuality in evolution. It greatly retards the final stages of multilocus adaptive change and severely limits the attainable precision of adaptation.

The only published work on the role of sex in evolution that takes account of the way meiosis destroys well adapted genotypes is that of Eshel and Feldman (1970). They concur with other workers that, given complete lack of epistatic effects on fitness, and a deterministic model of natural selection, asexual and sexual populations evolve at the same rate. When they introduce complex interactions between loci they conclude, in contradiction to the Mullerian theory, that sex can never accelerate adaptive evolution. If fitness of genes in combination is ever greater than the product of their separate fitnesses, sex retards evolution. Their analysis confirms for epistatic fitness interactions what was already obvious for heterozygote advantage. Sex generates recombinational load, and this largely annuls the effect of selection in each generation. They also show that mutational load will always be greater in sexual populations.

Advantages of asexual reproduction, often in combination with sexual, have often been recognized by students (e.g., Stebbins, 1970) of those higher plants in which both occur. A favorable genotype, once produced by mutation or recombination, can multiply without losing the combination and thereby bring all responsible genes to a high local frequency. Even if clonal multiplication is limited to a single season, the genes of a highly fit genotype can become locally more abundant than they could without asexual reproduction and increase the likelihood of similar combinations occurring again in the near future.

Preservation of favorable combinations within a genome can be achieved in sexual reproduction by having crossover rates lower than selection coefficients. Infrequent recombination can allow a given chromosome to exist mostly as Ab and aB if AB or ab is disadvantageous, whether the low recombination rate results from infrequent sexual reproduction or infrequent crossing over. The significance or linkage is discussed in Chapter 9. It is in single-locus segregational load that sexual reproduction must have a consistent disadvantage. If the ideal genotype at the a-locus is Aa, nearly all individuals produced asexually can inherit this combination, but only half (at most) of those produced sexually. Only a quarter of the sexually produced individuals can be AaBb, and so on.

There seems to be no escape from the conclusion that an asexual population should consist largely of one or a few highly fit genotypes, while the sexual population will be made up of a great variety of genotypes, of lower average fitness. This expectation is borne out by field data, reviewed below (pp. 160–162). Differences in genetic structure of an asexual and a sexual population should result in different responses to changed conditions, as suggested in Figure 15, which depicts

151

Before Environmental Change   After Environmental Change

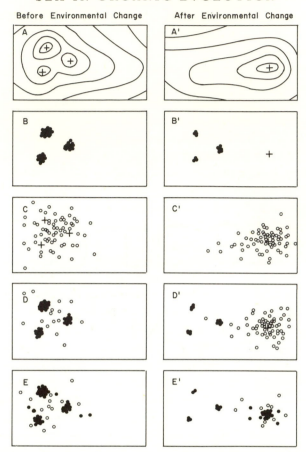

FIGURE 15. Effects of asexual and sexual reproduction on short-term evolutionary change. Initial states are shown to the left, those following an environmental change to the right. Members of clones shown as ●, exclusively sexual individuals as o. Ordinates and abscissas are multi-locus genotype functions chosen to maximize cluster resolution by two-dimensional projection. A-A′ shows levels of fitness, initially a general adaptive peak with three pinnacles well within the variability of a single population. B-B′ shows genotype distributions of clones incapable of reproducing sexually. C-C′ shows distribution for a Mendelian population, perhaps ancestral to the clones. D-D′ shows clones and Mendelian population in competition; occupation of the adaptive pinnacles by the clones greatly depresses

the proposed advantages and disadvantages of the two modes of reproduction. Asexual reproduction permits the fittest clone to appropriate a pinnacle of adaptation and locally win out over a Mendelian population. The Mendelian population, after an environmental change, can rapidly shift to at least the general neighborhood of the new pinnacles. If it has both asexual and sexual reproduction, a population can enjoy both advantages. The concept of a general adaptive peak being made up of a complex of pinnacles and depressions has some factual support from studies of Drosophila (Band, 1972). The illustrated numerical relations among clones and sexual populations has some support from data reviewed in the next chapter.

The model (Figure 15) includes a restatement of Maynard Smith's (1968A, 1971A) suggestion that sexual reproduction has its most positive effect when propagules from different sources colonize a new habitat. I would urge that the colonization process not be given too narrowly geographic an interpretation. In a heterogeneous and fluctuating environment, each new generation may be regarded as colonists entering new environments. If these colonists were sexually produced, they are more likely to include genotypes favored under the new conditions. If selection is intense, these temporarily favored individuals would have the sisyphean genotypes discussed in Chapter 5. Arguments on competition between asexual and sexual individuals in a population can be extrapolated to competition between asexual and sexual populations. Continued survival may depend on producing a diversity of genotypes at least

abundance of sexuals. E-E' shows the same competition, but with the original clones assumed capable of occasional sexual reproduction, which permits occupation of the new adaptive pinnacle. The original clones are severely depressed in numbers by the environmental change and consequent reduction in their fitness, and by competition from other genotypes of higher fitness, especially the new clone in E-E'.

153

as often as environmental change alters the relative fitness of genotypes.

CONCLUSIONS

The suggestion made here is that genotypic variety provides a margin of safety against environmental uncertainty, and that this may be more important to population survival than is precise adaptation to current conditions. Sexual reproduction facilitates evolution indirectly by making extinction less likely, not by making phylogenetic change more rapid. A surviving population may ultimately give rise to a great array of new taxa. An extinct one obviously cannot. As pointed out at the start of the chapter, the idea that genotypic variety is of long-term benefit is prevalent in general works on evolution, but I believe that the reasoning behind it is often faulty. It may be based on a sound inference that recombination can occasionally produce a genotype of high fitness. It ignores the equally valid point that sexual reproduction just as readily destroys such genotypes.

Both points of view are in conflict with most of the careful theoretical work from Fisher (1930) to Maynard Smith (1971A) and Kimura and Ohta (1971). This work proposes that recombination facilitates gene substitution and thereby speeds phylogenetic change. This is especially explicit in the work of Kimura and Ohta.

Unfortunately the facts of biology, as reviewed in the next chapter, offer little encouragement on any of these three theories. I would perhaps claim that the information is somewhat less embarassing for my own theory than for either of the others. It may be noted here that all three theories assume that sex is, in some way, of long-range benefit. As the next chapter will attempt to show, even this assumption may be regarded with suspicion.

154

CHAPTER THIRTEEN

Sex as a Factor in Biotic Evolution

There are two kinds of evolution of interest to biologists (Williams, 1966A). Organic evolution is change in the genetic constitution of a population. Biotic evolution is change in the composition of a biota. If a population evolves by organic evolution, the biota to which it belongs also evolves, but biotic evolution need not imply organic evolution. A succession of species in an abandoned field, an extirpation or successful invasion, any change in relative abundance of species would all be biotic evolution. They can all occur without genetic change in the constituent populations.

The kind of biotic evolution of primary concern in this chapter is the origin and extinction of taxa on the Earth as a whole over considerable lengths of geologic time. The main question to be considered is: Does the presence or absence of sexual reproduction in different taxa influence biotic evolution by altering rates of extinction, and if so, how?

Before considering the specific role of sexuality a more general discussion of extinction and biotic evolution is necessary. Extinction, especially of higher taxa, is obviously of great importance both as cause and effect in biotic evolution. Organic evolution during the Mesozoic produced an array of adaptations that constituted the dinosaur-way-of-life. Because of extinction these many adaptations are now missing from the biota. Their absence was undoubtedly a major contribution to the proliferation of mammalian adaptations.

General acceptance of the importance of extinction still permits a wide range of opinion on the nature of that importance. There are those who propose that extinction of local populations is so frequent and strongly biased in relation to different evolved adaptations that it can make a species evolve in a direction opposite to that of selection within each population. This is the theory of group selection, advocated mainly by those who see in social and reproductive processes a kind of adaptation that could not be produced by individual advantage (Wynne-Edwards, 1962; see also Simberloff and Wilson, 1969, and Boorman and Levitt, 1972). Avowed advocates of this theory are few, but casual group-selectionist thinking is common. Anyone who maintains that sex was evolved and is maintained because it confers long-range group advantage is willy-nilly, an advocate of group selection.

Many biologists who reject group selection as a major evolutionary force might support what could be called a theory of biased extinction. Given that two taxa have evolved different sets of adaptations, it seems reasonable to assume that the two complexes of characters may be unequally vulnerable to extinction. Obviously this force of extinction was heavily biased against the adaptations of several reptilian orders in the late Mesozoic.

Or was it? This chapter will argue that there is little respectable evidence that extinction has any consistent biases. It will propose that a taxonomically random force of extinction can produce phylogenies that may seem to be heavily favoring certain developments over others. It will contend that consistently recognizable causes of extinction are seldom, perhaps never demonstrable. It suggests that the usual textbook explanations for the extinction of this group and the proliferation of that may be of some pedagogic value in dramatizing the succession of taxa seen in the fossil record, but that they are part of the folklore of scientists, not science. More specifically, it will consider whether extinction shows any bias in relation to the incidence of sexual reproduction in life cycles.

ADAPTATION, EXTINCTION, AND GROUP SELECTION

As was clearly shown by Eiseley (1959 and references therein) the idea of natural selection is an old one. Many before Darwin grasped and correctly applied the idea. That no one before Darwin formulated natural selection as a basic principle shows, not that they doubted that the process was real, but that they failed to appreciate its importance. Given the observed degree of species stability and the assumption that human history in document and legend comprises all or most of Earth's history, natural selection must be a force of minor importance. Likewise after Darwin, all sane controversy centered on the adequacy of natural selection to produce the effects attributed to it, not on the reality of the process.

Similarly a profitable discussion of group selection must relate to the question of importance, not of reality. No one is likely to deny, as implied by Levins (1970), that a population's gene frequencies may have some effect on its persistence. This does not justify Levins' position that if group selection is real it deserves attention from a busy biological community. To deserve attention it needs some minimal degree of importance. The problem is not: Is it there? but rather: Has it produced important effects?

For group selection to be a major factor in evolution it is not enough that extinctions have causes; the causes must act consistently. The passenger pigeon might have survived with a different set of adaptations, perhaps something like those of the English sparrow. The interesting question is whether this comparison illustrates a basis for predicting survival and extinction. The pigeon was large, tree-nesting, and had a low potential rate of increase. The sparrow was smaller, ledge-nesting, and had a high potential rate of increase. Can we use these characters elsewhere to predict survival or extinction? Obviously not; there are too many apparently successful birds that are even larger, more arboreal, and less prolific than the passenger pigeon.

In the preceding chapter I proposed that natural populations can always evolve as fast as is required of them and that failure to keep up with demands of selection is never a cause of extinction. This is not the common view, and some distinguished biologists hold otherwise. Simpson (1963:289) says that between environmental demand and evolutionary response there may be a ". . . lag (extremely common, the usual or universal cause of extinction)," and on page 295 he says, "To explain any particular case of extinction it is, then, necessary to specify two things: what pertinent change occurred in the environment, and what factors in the population prevented sufficient adaptive change."

This illustrates the fallacy of regarding adaptation, not as that by which an individual maximizes its genetic influence on the population to which it belongs, but as that which normally acts to prevent extinction. The word "sufficient" clearly indicates Simpson's belief that adaptive change is always in a direction appropriate to forestall extinction, but is not always fast enough. I would contend that while evolution is always fast enough, it may be inappropriate in relation to population survival.

The most obvious example would be a population that adapts more and more perfectly to a disappearing habitat. Perhaps the passenger pigeon had a newly acquired obligate parasite that was perfecting its parasitic specialization at an unusually rapid rate. Consider also the green turtle population that nests on Ascencion Island. It spends most of its life near American shores, but every two years the adults undertake a remarkable feat of navigation and locate the tiny breeding area in the middle of the Atlantic. I assume that this population is closely adapted to its ecological niche, which includes this geographically restricted breeding ground. If erosion prevails over constructive forces and the island disappears, the last turtle to lay her eggs there may still be extremely well adapted to her niche, but it is a niche that in the next generation will permit a total of zero occupants.

The disappearance of an ecological niche must be a far more general phenomenon than the destruction of a bit of geography. A population may even obliterate its own niche by becoming better adapted to it. All that is required is that increasing adaptation have a progressively adverse effect on total resources. The extirpation of a host by a dependent parasite is merely one example of exploiters adversely affecting a resource.

Andrewartha and Birch (1954) summarize a large amount of circumstantial evidence that local extirpations may be commonplace, especially in insects, and Skellam (1955) calculated that even with considerable density-dependent control, random fluctuations that would seem minor on a scale of years may make a random walk to zero almost inevitable on a scale of millenia. A high frequency of local extirpations and taxonomic extinctions is necessary to a theory of group selection, but it is not sufficient. The extinctions must occur in groups that are consistently different from those that survive.

Various consistent reasons for extinction have been proposed. Various groups became extinct because they were overspecialized. This is undoubtedly true in the tautological sense that the more stringent the requirements for survival, the less likely they are to be met. The disappearance of gymnosperm taxa is attributed either to a supposed inferiority of their vascular systems or to their exposed seeds. The supposedly inferior wing support of pterodactyls (compared to bats) is implicated in their extinction. This sort of speculation is empty of scientific meaning, because there is seldom any real evidence for the inferiority of the adaptations in question, and never any for its role in extinction. Even if such evidence were produced, it may be inadequate support for general conclusions on bias in extinction. Single-finger wing support may have been an unfortunate committment for the pterodactyls, but could be nearly ideal for some other group. Having several fingers free for some other functions could preadapt a group to important evolutionary developments.

On my view extinction overtakes organisms that have about as close an approximation to evolutionary equilibrium in their

159

gene frequencies as those that survive. Extinction occurs because there is no corrective feedback between dangerously low population size and the forces of evolution. The last pair of passenger pigeons to nest successfully had no way of knowing that they ought to take desperate actions. Approaching extinction evokes no emergency measures, but rather ". . . the species doomed to extinction, innocently unconscious of its lack of 'fitness,' continues happily to perform its traditional rites" (Green, 1969).

GENETIC POTENTIAL AND ECOLOGICAL PERFORMANCE OF SPECIES WITH RESTRICTED RECOMBINATION

It is generally believed that sexual reproduction is a positive factor in the survival and evolutionary potential of a population. If this beneficial effect is important, a relative inferiority of asexual forms, in genetic potential and ecological performance, ought to be demonstrable. Observations offer little support for this expectation. Obviously there is no support from micro-organisms, in most of which sexual reproduction is rare and in many of which it is absent. The enormous population sizes of micro-organisms assure that great diversity, merely from mutation pressure, will always be available. Response-to-selection experiments strongly confirm this view.

Studies of purely asexual species of higher organisms also show that genetic diversity may be great (White, 1970; Solbrig, 1971; Suomalainen, 1962, 1969). Species that normally reproduce by self-fertilization will consist largely of homozygotes but different individuals may be homozygous for different alleles, and the population may hold a great store of genetic variability. This is evident from studies by Kannenberg and Allard (1967), Clegg, Allard, and Kahler (1972), and Jain and Marshall (1967) on self-fertilizing plant species. The populations studied were genetically diverse, showed higher

160

than expected levels of heterozygosity, and would obviously be capable of rapid response to selection.

If "evolutionary potential" means anything that can be related to observation, it must mean an ability to survive environmental change and exploit new opportunities. Available information offers little support for the view that frequent recombination increases evolutionary potential. Allard (1965) generalizes that it is the self-pollinating plants that are especially successful at colonizing new habitats. Their initial establishment must certainly be aided by the ability of a single seed to establish a colony, but thereafter the colony must be able to survive whatever conditions prevail in the new area. Having reproduction contingent on opposite-sex pairs may also decrease likelihood of continued survival. Skellam (1955) and Klomp, van Montford, and Tammes (1964) show that local extirpation is more likely for a dioecious species than for one that can reproduce asexually or by self-fertilization.

The versatility of exclusively apomictic plants is impressively demonstrated by the dandelion (Solbrig, 1971). I have seen them growing luxuriantly in both Florida and Iceland. Like other weeds, the dandelion is largely confined to cultivated or otherwise novel habitats. Asexuality and selfing are common among weeds (Allard, 1965; and other works in the same volume), even among animal "weeds" such as parthenogenetic lizards of the genus *Cnemidophorus* (Wright and Lowe, 1968). It is not at all evident whether the success of asexual or selfed weeds results from their ability to establish themselves with single propagules, their lack of reproductively wasteful male functions, their frequent polyploidy, or freedom from recombinational load.

Asexual species often out perform their sexual relatives. Suomalainen (1969) found that species groups of weevils often consist of one sexual form and one or more parthenogenetic. The sexual member is usually of restricted geographic range and the parthenogenetic forms widespread and more numerous. It is unreasonable to attribute this entirely to the ability of a parthenogen to establish itself from a single colonist. Both the asexual and sexual forms may occur in the midst of a continuous land mass where each seems to have spread out to the limits of its environmental tolerance. The parthenogenetic forms, having spread more widely, must be more versatile in their requirements than the sexuals. That they can survive in a broader range of conditions now means that they could better withstand an environmental change within their present ranges.

Other examples of the seemingly greater evolutionary potential of asexual forms, as judged from current ecological and biogeographic observations, are found in psocid insects (Mockford, 1971) and hawkweeds (Stebbins, 1950). Parthenogenetic (often polyploid) forms are more widespread and abundant than closely related sexuals. In the psocids it is the sexual forms that are found mainly in unusual habitats, ecologically and geographically. The only possible examples that I have found of sexual forms being more widespread and numerous than related asexuals is in the lizard genus *Cnemidophorus* (Zweifel, 1965; Wright and Lowe, 1968).

EVIDENCE FROM TAXONOMIC DISTRIBUTION

An indirect but potentially decisive way of evaluating sex as an evolutionary factor is to look at its taxonomic distribution. The alternative character states *presence* and *absence* of sexual reproduction will be distributed differently among components of a biota according to various evolutionary considerations: rate of evolution from presence to absence, the reverse rate, and the rates of extinction and cladogenesis with and without sexual reproduction. As a hypothetical illustration, consider the character states *black* and *white* in Figure 16. If it could be shown that the phylogenetic distribution

162

PRESENT

FIGURE 16. Evolution of hypothetical character states black and white. At the left, black occasionally changes to white, but the reverse happens even more readily. In the center, black occasionally changes irreversibly to white, and this increases the danger of extinction. At the right, black changes to white only once, but here it makes extinction less likely. The data of systematics would not ordinarily distinguish the left from the center phylogeny.

of *absence* of sexuality were distributed much the same as *white* in the center dendrogram, there would be clear evidence of a positive effect of sex on population survival.

The claim that sex is of long-term advantage in evolution is often supported by an argument from taxonomic distribution (Mayr, 1963; Rollins, 1967; Stebbins, 1957, 1970; Weismann, 1889). Mayr, for example, states that exclusively asexual reproduction ". . . gains only a short-term advantage, and, with the apparent exception of the bdelloids, virtually every case of parthenogenesis in the animal kingdom has all the earmarks of recency." As should be clear from the figure, low taxonomic rank of groups with a certain character state could be used instead as an argument for the ready reversibility of that state. Only if it could be shown that loss of sexuality is irreversible would the low taxonomic rank of affected groups (Mayr's "earmarks of recency") be evidence of a long-term disadvantage in the loss of sexuality. It would also be desirable

163

to get some real data on the tendency of strictly asexual taxa to be of low rank.

It does seem plausible that complete loss of the machinery of sexual reproduction would be irreversible, but it may be that a return to sexuality is latent in many currently asexual populations. In a number of plants the absence of sexuality has a simple genetic basis that could be easily reversed by mutation or introgression and favorable selection for sexuality (Lewis, 1942). Sexually sterile triploids can be cured by altered ploidy. Gynogenetic fishes may become sexual if males of their own species are available.

It is also often true that the lack of sexuality in a species in marginal habitats has no genetic basis at all. It arises from environmental conditions that make it impossible to carry out some process essential to sexual reproduction. Remane and Schlieper (1971) describe many examples among plants and animals from the sea or fresh waters that have invaded brackish waters. There are seaweeds that form nonfunctional sporangia and higher plants that flower but set no seed. For some animal forms the abnormal salinity prevents fertilization and in others the larvae are unable to survive. Adult stages of all these organisms reproduce vegetatively and thereby maintain themselves in brackish water. Stress of abnormal salinity is undoubtedly only one of many factors that may interfere with the complex process of sexual reproduction in marginal habitats.

To be convincing, the argument from taxonomic rank will require a more thorough documentation than it has received. The pattern of branching and extinction in the two left dendrograms (Figure 16) was selected as of suitable gentral complexity from some purely random phylogenies (Figure 17). These were generated as follows. A high-rank taxon with ten subordinate taxa is represented by ten lines rising from the bottom of the graph. Any taxonomic level can be assumed, but for parallelism with textbook examples of extinction and

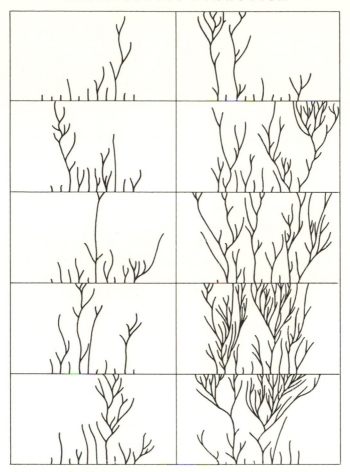

FIGURE 17. Random phylogenies, generated as explained in the text, and presented in order of increasing success of the group as a whole.

adaptive radiation I will speak of a class with ten orders. Extinction and branching are randomized. On the average one ancestral order is represented by one descendant order after one unit of time, but with chance extinction balanced, on the average, by chance cladogenesis. I assume that getting numbers of surviving orders from a Poisson distribution with $\lambda = 1$ is a realistic model. The figures show my first ten random phylogenies in order of increasing class success.

It is instructive to search the random phylogenies for sudden extinctions of once dominant groups, for which some catastrophe might be postulated; for more steady attritions, which might be attributed to some proposed inferiority; for adaptive radiations from ancestors that must obviously have had great evolutionary potential, recognizable perhaps in some adaptive innovation or "general adaptation"; and for such anomalies as "living fossils." It is also instructive to compare some of the random phylogenies with that used by Wright (1956) as evidence for the selectiveness of extinction of vertebrate taxa.

A realistic model of random cladogenesis and extinction would produce more variable phylogenies than those illustrated, because one of the constants ought really to be variable. Expectations of splitting or extinction should both change in time and vary among orders, to reflect variation in rates of environmental change. I have no suggestions on what would be a realistic pattern of variation in rates of environmental change.

The uniform time scale makes Figure 17 a conservative model of random variation in phylogenetic success and failure, but it suffices to illustrate some pitfalls of intuitive arguments on cause and effect in evolutionary history. Suppose the second class was asexual and the ninth sexual. One is now represented by a single "living fossil" and the other by 20 orders from 3 different ancestors. One might be tempted to attribute the greater success of the ninth to its sexual reproduction, but this conclusion would not be significant at the 0.05 level. Demonstration of consistent causes in phylogeny demands that seeming biases be tested in relation to null hypotheses based on realistic models of random phylogeny. Methods of numerical phylogeny as developed by Sneath and

Sokal (1973) may furnish some needed methodology. Branching theory should be as applicable to survival and extinction of taxa as to the survival and extinction of mutations (Schaffer, 1970).

Whether adequate phylogenetic data are available for testing for consistent bias in extinction and radiation is another problem. Possible bias in relation to presence or absence of sexual reproduction would certainly be one of the more important phylogenetic problems to be attacked. It is also, in my opinion, about the only one for which there is some likelihood of positive findings. It is conceivable that sex plays somewhat similar roles in the elm and the aphid that feeds on it. Such functional similarity seems unlikely for greenness or any other special feature that they may have in common.

CONCLUSIONS

It may be some time before a reliable measure of nonrandomness in the phylogenetic distribution of the loss of sexuality will appear. Meanwhile I am inclined to accept provisionally the intuitions of leading animal and plant systematists from Weismann to Mayr and Stebbins and assume that loss of sexuality really does reduce prospects for long-term survival and adaptive radiation. If this conclusion is correct, how do we reconcile it with evidence, reviewed above, that asexual and self-fertilized species seem so versatile in adapting to novel environments?

A possible resolution may be found in the mode of origin of asexual species. Stebbins (1950:107–109) suggests that the advantages of the asexual forms may derive, not from their own evolutionary processes, but from immediately ancestral sexual forms. Some sexual species may occasionally produce individuals able to reproduce asexually but not sexually. Very rarely, such a defective individual may have a sisyphean genotype in relation to local conditions. It will become locally

167

abundant and extirpate or severely depress the numbers of the ancestral sexual species. If this can happen in one locality, it can happen in another. Ultimately there will be two "species." The sexual one will survive in limited numbers in rare refuges where it is free from competition from its asexual derivatives. The asexual species will consist of locally dominant clones derived, not from each other, but from the rare sexual relative.

Collectively versatile asexual complexes may often originate in this way, but the process is hardly sufficient to account for all of the success of asexual species. Some have spread as weeds to diverse climatic regions far beyond the range of sexual ancestors. Their adaptability would seem neither locally limited nor temporary. Also, selfing may produce almost as clonal an inheritance as asexual reproduction, along with inbreeding depression. It must originate gradually from favorable selection in a whole population. The inbred complexes studied by Allard and collaborators (e.g. Allard, 1965; Clegg, Allard, and Kahler, 1972) are highly successful invaders of the novel habitats provided by man, but must have originated from each other, rather than independently from an outcrossed ancestor.

There are some other possible explanations, at best only partial. Occupation of a new habitat may depend more on the presence of uniparental reproductive capacities than on the absence of outcrossing. Normally outcrossed plants may persist vegetatively or apomictically if special pollen-vector requirements are met. Failure of sexual reproduction in brackish water was mentioned above. If novel habitats persist long enough, it may be that the success of the uniparentals will prove temporary. Better adapted outcrossed forms may evolve and crowd them out.

Also it may be that apparently asexual or selfed species outcross on rare occasions. Perhaps an extremely low rate of recombination can generate sufficient variety for rapid evolu-

tion. The situation may be that of Figure 15ᴇ, rather than
ʙ or ᴅ (page 152). Such species may be enjoying the best
of both the worlds of asexual and sexual reproduction, but
perhaps only briefly. Selection within the population may result in either the complete loss of outcrossing or its increase
to the point at which highly favorable gene combinations cannot persist for a significant period of time.

The possibility should also be considered that a high level
of adaptation in the great majority of individuals, as is attainable in a complex of specialized clones, even with occasional
outcrossing, is not the most favorable population structure on
a long-range evolutionary time scale. Eshel and Feldman
(1970) suggested that there may be some benefit in the capacity of sexual reproduction to retard organic evolution. They
described this retardation as a prolongation of the "polymorphic state." I see no reason why there should be more polymorphic loci or heterozygosity in a sexual species, and would
have been happier with the term "genotypic variety." Also,
Bateson (1963), in considering a hypothetical population with
Lamarckian inheritance, concluded that in the long run it
could not compete with a Mendelian population evolving by
natural selection. It would suffer from excessive responsiveness
to strong evolutionary demands and become overspecialized
for fleeting niches.

I am sure that many readers have already concluded that
I really do not understand the role of sex in either organic
or biotic evolution. At least I can claim, on the basis of the
conflicting views in the recent literature, the consolation of
abundant company. Clearly the contest of ideas on these fundamental problems has only just begun. History has afforded
a rare opportunity to ardent participants and alert spectators
in the years ahead.

Bibliography

Acton, A. B. 1961. An unsucessful attempt to reduce recombination by selection. *Amer. Naturalist,* 95:119–120.

Allard, R. W. 1965. Genetic systems associated with colonizing ability in predominantly self-pollinating species. In: *Genetics of Colonizing Species* (pp. 49–76), H. G. Baker and G. L. Stebbins, editors, New York, Academic, *xv* + 588 pp.

Andrew, R. J. 1966. Precocious adult behavior in the young chick. *Animal Behavior,* 14:485–500.

Andrewartha, H. G. and L. C. Birch. 1954. *The distribution and abundance of animals.* Univ. Chicago Press, *xv* + 782 pp.

Antonovics, Janis, and A. D. Bradshaw. 1970. Evolution in closely adjacent plant populations. VIII. Clinal patterns at a mine boundary. *Heredity,* 25:349–362.

———, ———, and R. G. Turner. 1971. Heavy metal tolerance in plants. *Adv. Ecological Research,* 7:2–85.

Apfelbach, Raimund. 1966. Maulbruten und Paarbindung bei *Tilapia galilaea* L. (Pisces, Cichlidae). *Die Naturwissenschaften,* 53:22.

Applegate, Vernon C. 1951. Sea lamprey investigations. II. Egg development, maturity, egg production, and percentage of unspawned eggs of sea lampreys, *Petromyzon marinus,* captured in several Lake Huron tributaries. *Papers Michigan Acad. Sci.,* 35:71–90.

Aronson, Lester R. 1948. Problems in the behavior and physiology of a species of African mouthbreeding fish. *Trans. New York Acad. Sci.* (2)2:33–42.

Asher, James H., Jr. 1970. Pathenogenesis and genetic variability. II. One-locus models for various diploid populations. *Genetics,* 66:369–391.

Asker, Sven. 1970. Apomictic biotypes in *Potentilla intermedia* and *P. norvegica. Hereditas,* 66:101–108.

170

————. 1971. Apomixis and sexuality in the *Potentilla argentea* complex. III. Euploid and aneuploid derivatives (including trisomics) of some apomictic biotypes. *Ibid.,* 67:111–142.

Bacci, Guido. 1965. *Sex determination.* New York, Pergamon, 306 pp.

Baer, Jean G. 1952. *Ecology of animal parasites.* Urbana, Univ. Illinois Press, *x* + 224 pp.

Baerends, G. P. and J. M. Baerends-VanRoon. 1950. An introduction to the study of the ethology of cichlid fishes. *Behaviour* (Suppl. 1) : 1–243.

Bagenal, T. B. 1957. The breeding and fecundity of the long rough dab, *Hippoglossoides platessoides* (Fabr.) and the associated cycle in condition. *J. Mar. Biol. Assn., U.K.,* 36:339–375.

————. 1963. The fecundity of plaice from the Bay of Biscay. *Ibid.,* 43:177–179.

————. 1965. The fecundity of long rough dabs in the Clyde Sea area. *Ibid.,* 45:599–606.

————. 1966. The ecological and geographical aspects of the fecundity of the plaice. *Ibid.,* 46:161–186.

Baker, R. J. 1969. Genotype-environment interactions in yield of wheat. *Canadian J. Plant Science,* 49:743–751.

Band, H. T. 1972. Minor climatic shifts and genetic changes in a natural population of *Drosophila melanogaster. Amer. Naturalist,* 106:102–115.

Bannister, M. H. 1965. Variation in the breeding system of *Pinus radiata.* In: *The Genetics of Colonizing Species* (pp. 353–374), H. G. Baker and G. L. Stebbins, editors, New York, Academic, *xv* + 588 pp.

Barber, H. N. and W. D. Jackson, 1957. Natural selection in action in Eucalyptus. *Nature,* 179:1267–1269.

Barlow, George W. and Richard F. Green. 1970. The problems of appeasement and of sexual roles in the courtship

171

behavior of the blackchin mouthbreeder, *Tilapia melano-teron* (Pisces: Cichlidae). *Behaviour*, 36:84–115.

Bateman, A. J. 1949. Analysis of data on sexual isolation. *Evolution*, 3:174–177.

Bateson, Gregory. 1963. The role of somatic change in evolution. *Evolution*, 17:529–539.

Beach, Frank A. 1964. Biological bases for reproductive behavior. In: *Social Behavior and Organization among Vertebrates* (pp. 117–142), W. Etkin, editor, Chicago, Univ. Chicago Press, xii + 307 pp.

Beardmore, John A. 1963. Mutual facilitation and the fitness of polymorphic populations. *Amer. Naturalist*, 97:69–74.

Beatty, R. A. 1957. *Parthenogenesis and polyploidy in mammalian development*. Cambridge, Cambridge Univ. Press, xi + 132 pp.

———. 1967. Parthenogenesis in vertebrates. In: *Fertilization: Comparative morphology, biochemistry, and immunology*, vol. 1 (pp. 413–440) C. B. Metz and A. Monroy, editors, New York, Academic, xiii + 489 pp.

Beverton, R. J. H. 1962. Long-term dynamics of certain North Sea fish populations. *British Ecol. Symp.*, 2:242–259.

——— and S. J. Holt. 1957. On the dynamics of exploited fish populations. *Fishery Investigations* (London) (2), 19:1–533.

Birky, C. William, Jr. and John J. Gilbert. 1971. Parthenogenesis in rotifers: The control of sexual and asexual reproduction. *Amer. Zoologist*, 11:245–266.

Blaxter, J. H. S. 1965. The feeding of herring larvae and their ecology in relation to feeding. *Calif. Coop. Oceanic Fish. Invest., Repts.*, 10:79–88.

Bliss, C. I. and K. A. Reinker, 1964. A lognormal approach to diameter distributions in even-aged stands. *Forest Sci.*, 10:350–360.

Bodmer, Walter F. 1965. Differential fertility in population genetics models. *Genetics*, 51:411–424.

———. 1970. The evolutionary significance of recombination

in prokaryotes. *Symp. Soc. General Microbiol.*, 20:279–294.

———. 1972. The evolution of recombination mechanisms in bacteria. In: *Uptake of informative molecules by living cells* (pp. 130–140), ed. L. Ledoux, Amsterdam, North-Holland Publ. Co., xi + 416 pp.

Bonner, John T. 1958. The relation of spore formation to recombination. *Amer. Naturalist*, 92:193–200.

Boorman, Scott, A. and Paul R. Levitt. 1972. Group selection on the boundary of a stable population. *Proc. Nat. Acad. Sci.*, 69:2711–2713.

Bourliere, Francois. 1964. *The natural history of mammals.* New York, Knopf, *xxi* + 387 + *xi*.

Boyden, Alan A. 1954. Comparative evolution with special reference to primitive mechanisms. *Evolution,* 7:21–30.

Brachet, J. and P. Malpoix. 1971. Macromolecular synthesis and nucleocytoplasmic interactions in early development. *Adv. Morphogenesis,* 9:263–316.

Breder, Charles M., Jr. 1935. The reproductive habits of the common catfish, *Ameiurus nebulosus* (LeSueur), with a discussion of their significance in ontogeny and phylogeny. *Zoologica,* 19:143–185.

——— and Donn Eric Rosen. 1966. *Modes of reproduction in fishes.* Garden City, N.J., Natural History Press, xv + 941 pp.

Brown, Edward H., Jr. 1972. Population biology of alewives, *Alosa pseudoharengus* in Lake Michigan, 1949–70. *J. Fish. Res. Board Canada,* 29:477–500.

Cable, Raymond M. 1971. Parthenogenesis in parasitic helminths. *Amer. Zoologist,* 11:267–272.

Carson, Hampton L. 1967. Selection for parthenogenesis in *Drosophila mercatorum. Genetics,* 55:157–171.

Cheng, Thomas C. 1964. *Biology of animal parasites.* Philadelphia, Saunders, 727 pp.

Chinnici, Joseph P. 1971. Modification of recombination frequency in Drosophila. i. Selection for increased and decreased crossing over. *Genetics,* 69:71–83.

1 7 3

Chittleborough, R. G. and L. R. Thomas. 1969. Larval ecology of the western Australian marine crayfish, with notes upon other panulirid larvae from the eastern Indian Ocean. *Australian J. Marine and Freshw. Res.,* 20:199–223.

Clark, Frances N. and John C. Marr. 1955. Population dynamics of the Pacific sardine. *Calif. Coop Oceanic Fish. Invest., Rept.,* 1953–55:11–48.

Clarke, C. A., C. G. C. Dickson, and P. M. Sheppard. 1963. Larval color pattern in *Papilio demodocus. Evolution,* 17:130–137.

Clausen, Jens. 1954. Partial apomixis as an equilibrium system in evolution. *Caryologia,* 6(Suppl.):469–479.

—— and W. M. Heisey. 1958. Experimental studies on the nature of species. IV. Genetic structure of ecological races. *Carnegie Inst. Washington Publ.,* 615:1–312.

Clegg, M. T., R. W. Allard, A. L. Kahler. 1972. Is the gene the unit of selection? Evidence from two experimental plant populations. *Proc. Nat. Acad. Sci.,* 69:2474–2478.

Coe, Wesley R. 1953. Resurgent populations of littoral marine invertebrates and their dependence on ocean currents and tidal currents. *Ecology,* 34:225–229.

Cohen, Dan. 1968. A general model of optimal reproduction in a randomly varying environment. *J. Ecology,* 56:219–228.

Cook, L. M. 1971. *Coefficients of natural selection.* London, Hutchinson Univ. Libr., 207 pp.

Cook, Sheila C. A., Claude Lefevre, Thomas McNeilly. 1972. Competition between metal tolerant and normal plant populations on normal soil. *Evolution,* 26:366–372.

Crawford-Sidebotham, T. J. 1972. The role of slugs and snails in the maintenance of the cyanogenesis polymorphisms of *Lotus corniculatus* and *Trifolium repens. Heredity,* 28:405–411.

Crew, F. A. E. 1965. *Sex determination.* London, Methuen, *viii* + 188 pp.

Crow, James F. 1968. The cost of evolution and genetic loads.

In:*Haldane and Modern Biology* (165–178), K. R. Drona-
mraju, editor. Baltimore, Johns Hopkins, *xvi* + 333 pp.

———— and Motoo Kimura. 1965. Evolution in sexual and
asexual populations. *Amer. Naturalist,* 99:439–450.

———— and ————. 1969. Evolution in sexual and asexual
populations, *ibid.,* 103:89–91.

———— and ———— 1970. *An introduction to population genetics theory.* New York, Harper and Rowe, *xv* + 591 pp.

Crumpacker, D. W. 1967. Genetic loads in maise (*Zea mays*
L.) and other crossfertilized plants and animals. *Evolutionary Biol.,* 1:306–424.

Cushing, D. H. 1971. The dependence of recruitment on
parent stock in different groups of fishes. *J. du Conseil
Intern. Expl. Mer,* 33:340–362.

Dahne, C, 1914. *Alytes obstetricans* und seine Brutpflege. *Bl.
Aquarienkunde* (Stuttgart), 25:227–229.

Darwin, Charles R. 1871. *The descent of man, and selection
in relation to sex.* New York, Appleton, vol. 1, *vi* + 409
pp., vol. 2, *vi* + 436 pp.

Davidoff, Edwin B. 1965. Estimation of year class abundance
and mortality of yellowfin tuna in the eastern tropical
Pacific. *Bull. Inter-Amer. Tropical Tuna Comm.,* 10:355–
380.

Dobzhansky, Theodosius. 1964A. How do the genetic loads
affect the fitness of their carriers in Drosophila populations?
Amer. Naturalist, 98:151–166.

————. 1964B. Genetics of natural populations. xxxv. A
progress report on genetic changes in populations of *Drosophila pseudoobscura* in the American southwest. *Evolution,* 18:164–176.

Dodson, Edward O. 1953. Comments on the origin of sex and
of meiosis. *Evolution,* 7:387–388.

Downhower, Jerry F. and Kenneth B. Armitage. 1971. The
yellow-bellied marmot and the evolution of polygamy.
Amer. Naturalist, 105:355–370.

Drilhon, A., J. M. Fine, G. A. Boffa, P. Amouth. 1966. Les groupes de transferrines chez l'anguille. Differences phénotypicues entre les anguilles de l'Atlantique et les anguilles mediterranées. *C. R. Hebd. Séances Acad. Sci.* (D) 262:1315–1318.

————. 1967. Les groupes de transferrines chez *Anguilla anguilla*. Etude des deux populations d'origine geographique differente. *Ibid.,* 265:1096–8.

Edwards, A. W. F. 1967. Fundamental theorem of natural selection. *Nature,* 215:537–538.

Ege, Vilh. 1942. A transplantation experiment with *Zoarces viviparus* L. *C. R. Laboratory Carlsberg (ser. Physiol.),* 23:271–386.

Ehrensvärd, Gösta. 1962. *Life: Its origin and development.* Minneapolis, Burgess, 204 pp.

Ehrlich, Paul R. and Peter H. Raven. 1969. Differentiation of populations. *Science,* 165:1228–1232.

Eiseley, Loren C. 1959. Charles Darwin, Edward Blyth, and the thory of natural selection. *Proc. Amer. Philosoph. Soc.,* 103:94–158.

Elrod, Joseph H. 1969. Estimates of some vital statistics of northern pike, walleye, and sauger populations in Lake Sharpe, South Dakota. *Techn. Paper Bur. Sport Fish. and Wildl., U.S.,* 30:1–17.

Emlen, John M. 1966. Natural selection and human behavior. *J. Theoretical Biol.,* 12:410–418.

————. 1973. *Ecology: An evolutionary approach.* Reading, Mass., Addison-Wesley, *xiv* + 493 pp.

Epling, Carl, Harlan Lewis, Francis M. Ball. 1960. The breeding group and seed storage: A study in population dynamics. *Evolution,* 14:238–255.

Eshel, Ilan, and Marcus W. Feldman. 1970. On the evolutionary effect of recombination. *Theoretical Population Biol.,* 1:88–100.

176

Evans, A. Murray. 1969. Problems of apomixis and the treatment of agamic complexes. *BioScience,* 19:708–711.

Faegri, K. and L. van der Pijl. 1966. *The principles of pollination biology.* London, Pergamon, *ix* + 248 pp.

Faugères, A., C. Petit, E. Thibout. 1971. The components of sexual selection. *Evolution,* 25:265–275.

Fiedler, Kurt. 1955. Vergleichende Verhaltenstudien an Seenadeln, Schlangennadeln und Seepferdchen. *Zeitschr. f. Tierpsychol.,* 11:358–416.

Fisher, Ronald A. 1930. *The genetical theory of natural selection.* New York, Dover Reprint (1958). *xiv* + 291.

Frank, Peter W., Catherine D. Boll, Robert W. Kelly. 1954. Vital statistics of laboratory cultures of *Daphnia pulex* DeGeer as related to density. *Physiol. Zool.,* 30:287–305.

Franklin, E. C. 1972. Genetic load in the loblolly pine. *Amer. Naturalist,* 106:262–265.

Freeman, G. H. and Jean M. Perkins. 1971. Environment and genotype-environmental components of variability. viii. Relations between genotypes grown in different environments and measures of these environments. *Heredity,* 27:15–23.

Fripp, Yvonne J. and C. E. Caten. 1971. Genotype-environmental interactions in *Schizophyllum commune. Heredity,* 27:393–407.

Fritsch, Felix E. 1935. *The structure and reproduction of the algae,* vol. 1. Cambridge, Cambridge Univ. Press., *xvii* + 791 pp.

Fujino, Kazo, and Tagay Kang, 1968. Transferrin groups of tunas. *Genetics,* 59:79–91.

Gadgil, Madhav and William H. Bossert. 1970. Life history consequences of natural selection. *Amer. Naturalist,* 104:1–24.

Galtsoff, Paul S. 1964. The American oyster, *Crassostrea virginica* Gmelin. *Fish. Bull., U.S.,* 64:1–480.

BIBLIOGRAPHY

Gause, G. F. 1934. *The struggle for existence.* Baltimore, Williams and Wilkins, *ix* + 163 pp.

Ghiselin, Michael T. 1969. The evolution of hermaphroditism among animals. *Quart. Rev. Biol.,* 44:189–208.

Giesel, James T. 1972. Sex ratio, rate of evolution, and environmental heterogeneity. *Amer. Naturalist,* 106:380–387.

Gill, Douglas E. 1972. Intrinsic rates of increase, saturation densities, and competitive ability. I. An experiment with Paramecium. *Amer. Naturalist,* 106:461–471.

Graham, Michael. 1956. *Sea fisheries. Their investigation in the United Kingdom.* London, Edward Small, *xii* + 487 pp.

Grant, Verne, and Karen A. Grant. 1971. Dynamics of clonal microspecies in cholla cactus. *Evolution,* 25:144–55.

Grassle, J. Frederick. 1971. Species diversity, genetic variability and environmental uncertainty. In: *Fifth European Symposium on Marine Biology* (pp. 19–26), Bruno Battaglia, editor, Padua, Piccin Editore, *x* + 348 pp.

Grene, Marjorie. 1969. Bohm's metaphysics and biology. In: *Towards a Theoretical Biology,* part 2 (pp. 61–69), C. H. Waddington, editor, Chicago, Aldine, 351 pp.

Gulland, J. A. 1971. Ecological aspects of fishery research. *Adv. Ecological Res.,* 7:115–176.

Gunston, Gary K. 1968. Gonad weight/body weight ratio of mature chinook salmon males as a measure of gonad size. *Progressive Fish Cult.,* 30:23–25.

Gustafsson, Åke. 1946. Apomixis in higher plants. *Lunds Univ. Arsskrift.* N.F. (Avd. 2), 42:1–370.

Hamilton, W. D. 1964A. The genetical evolution of social behavior. I. *J. Theoretical Biol.* 7:1–16.

———. 1964B. The genetical evolution of social behavior. II. *Ibid.,* 7:17–89.

———. 1966. The moulding of senescence by natural selection. *Ibid.,* 12:12–45.

———. 1967. Extraordinary sex ratios. *Science,* 156:477–488.

178

BIBLIOGRAPHY

Harberd, D. J. 1967. Observations on natural clones in *Holcus mollis*. *New Phytologist,* 66:401–408.

———— and M. Owen. 1969. Some experimental observations on clone structure of a natural population of *Festuca rubra L., ibid.,* 68:93–104.

Harden Jones, F. R. 1968. *Fish migration.* London, Edward Arnold, *viii* + 325 pp.

Harper, John L. 1965A. Establishment, aggression, and cohabitation in weedy species. In: *Genetics of colonizing species* (pp. 243–268), H. G. Baker and G. L. Stebbins, editors, New York, Academic, *xv* + 588 pp.

————. 1965B. The nature and consequence of interference amongst plants. *Proc. 11th Intern. Congress Genetics,* 465–482.

————. 1966. The reproductive biology of the British poppies. In: *Reproductive biology and taxonomy of vascular plants* (pp. 26–39), J. G. Hawkes, editor, New York, Pergamon, 183 pp.

Haskell, Gordon. 1966. The history, taxonomy and breeding system of apomictic British Rubi. In: *Reproductive biology and taxonomy of vascular plants* (pp. 141–151), J. G. Hawkes, editor, New York, Pergamon, 183 pp.

Hawes, R. S. J. 1963. The emergence of asexuality in protozoa. *Quart. Rev. Biol.,* 38:234–242.

Healey, M. C. 1971. Gonad development and fecundity of the sand goby. *Gobius minutus* Pallas. Trans. Amer. Fisheries Soc., 100:520–526.

Heard, William R. 1972. Spawning behavior of pink salmon on an artificial redd. *Trans. Amer. Fish. Soc.,* 101:276–283.

Hebert, P. D. N. and R. D. Ward. 1972. Inheritance during parthenogenesis in *Daphnia magna. Genetics,* 71:639–642.

Hedgpeth, Joel W. 1963. Pycnogonida. *Encyclopedia Britannica,* 18:788.

Herald, Earl S. 1961. *Living fishes of the world.* New York, Doubleday, 304 pp.

179

BIBLIOGRAPHY

Hett, Joan M. 1971. A dynamic analysis of age in sugar maple seedlings. *Ecology*, 52:1071–1074.

——— and Orie L. Loucks. 1968. Application of life-table analysis to tree seedlings in Quetico Provincial Park, Ontario. *Forestry Chron.*, 44 (2):29–32.

———, ———. 1971. Sugar maple (*Acer saccharum* Marsh) seedling mortality. *J. Ecology*, 59:507–520.

Hodder, V. M. 1963. Fecundity of Grand Bank haddock. *J. Fish. Res. Board Canada*, 20:1465–1487.

Hoyt, Robert D. 1971. The reproductive biology of the silver-jaw minnow, *Ericymba buccata* Cope, in Kentucky. *Trans. Amer. Fish. Soc.*, 100:510–519.

Hyman, L. H. 1967. *The invertebrates. vi. Mollusca i.* New York, McGraw-Hill, *vii* + 792 pp.

Jackson, James F. 1970. Lognormal scale counts. *Systematic Zool.*, 19:194–196.

Jain, S. K. and D. R. Marshall. 1967. Population studies in predominantly self-pollinating species. x. Variation in natural populations of *Avena fatua* and *A. barbata*. *Amer. Naturalist*, 101:19–33.

Jensen, A. S. 1944. On specific constancy and segregation into races in sea-fishes. *Biologiske Meddeleser*, Copenhagen, 19(8):1–19.

Jinks, J. L. and Jean M. Perkins. 1970. Detection and estimation of genotype-environmental, linkage, and epistatic components of variation for a metrical trait. *Heredity*, 25:157–177.

Johnson, M. S. 1971. Adaptive lactate dehydrogenase variation in the crested blenny, *Anoplarchus*. *Heredity*, 27:205–226.

Jónsson, Jón. 1957. Thorskur. In: *Íslensk Dýr. i. Friskarnir*, by Bjarni Saemundsson (pp. 515–524). Reykjavík, Bókaverzlun Sigfúsar Eymundssonar, 583 pp.

Kalmus, H. and C. A. B. Smith. 1960. Evolutionary origin

of sexual differentiation and the sex-ratio. *Nature*, 186:1004–1006.

Kannenberg, L. W. and R. W. Allard. 1967. Population studies in predominantly self-pollinated species. VIII. Genetic variability in the *Festuca microstachys* complex. *Evolution*, 21:227–240.

Kaya, Calvin M. and Arthur D. Hasler. 1972. Photoperiod and temperature effects on the gonads of green sunfish, *Lepomis cyanellus* (Rafinesque), during the quiescent, winter phase of its annual sexual cycle. *Trans. Amer. Fish. Soc.*, 101:270–275.

Kendeigh, S. Charles. 1952. Parental care and its evolution in birds. *Illinois Biol. Monogr.*, 22:1–356.

Kerfoot, W. Charles. 1969. Selection of an appropriate index for the study of the variability of lizard and snake body scale counts. *Systematic Zool.*, 18:53–62.

Kidwell, Margaret Gale. 1972. Genetic change of recombination value in *Drosophila melanogaster*. I. Artificial selection for high and low recombination and some properties of the recombination-modifying genes. *Genetics*, 70:419–432.

Kimura, Motoo. 1967. On the evolutionary adjustment of spontaneous mutation rates. *Genetics Res.* (Cambridge), 9:23–34.

——— and Tomoko Ohta. 1971. *Theoretical aspects of population genetics*. Princeton, Princeton Univ. Press, ix + 219 pp.

King, Charles E. 1972. Adaptation of rotifers to seasonal variation. *Ecology*, 53:408–418.

King, Jack Lester. 1967. Continuously distributed factors affecting fitness. *Genetics*, 55:483–492.

Klekowski, Edward J., Jr. 1972. Evidence against genetic self-incompatability in the homosporous fern *Pteridium aquilinum*. *Evolution*, 26:66–73.

Klomp, H., M. A. J. van Monfort, P. M. L. Tammes. 1964.

181

Sexual reproduction and underpopulation. *Arch. Neerl. Zool.*, 16:105–110.

Koehn, Richard K. 1972. Genetic variation in the eel: A critique. *Marine Biology*, 14:179–181.

———— and Jeffry B. Mitton. 1972. Population genetics of marine pelecypods. i. Ecological heterogeneity and evolutionary strategy at an enzyme locus. *Amer. Naturalist*, 106:47–56.

————, Francis J. Turano, Jeffry B. Mitton. 1973. Population genetics of marine pelecypods. ii. Genetic differences in microhabitats of *Modiolus demissus. Evolution*, 27:100–105.

Kosuda, Kazuhiko and Daigoro Moriwaki. 1971. Increase of genetic variability through recombination in *Drosophila melanogaster. Genetics*, 67:287–304.

Koyama, Hirosi and Tatuo Kira. 1956. Intraspecific competition among higher plants. viii. Frequency distribution of individual plant weight as affected by the interaction between plants. *J. Inst. Polytech. Osaka City Univ.*, (d) 7:73–94.

Krumholz, Louis A. 1958. Relative weights of some viscera in the Atlantic marlins. *Bull. Amer. Mus. Natural Hist.*, 114:402–405.

————. 1959. Stomach contents and organ weights of some bluefin tuna, *Thunnus thynnus* (Linnaeus), near Bimini, Bahamas. *Zoologica*, 44:127–131.

Kudo, Richard R. 1966. *Protozoology.* Springfield, Thomas, xi + 1174 pp.

Lack, David. 1954. *The natural regulation of animal numbers.* Oxford, Oxford Univ. Press, viii + 343 pp.

————. 1968. *Ecological adaptations for breeding in birds.* London, Methuen (New York, Barns and Noble), xii + 409 pp.

Lancaster, Douglas A. 1964. Biology of the brushland tinamou, *Nothoprocta cinerascens. Bull. Amer. Mus. Natural Hist.*, 127:271–314, pls. 16–26.

Lannon, James E. 1971. Experimental self-fertilization of the

Pacific oyster, *Crassostrea gigas,* utilizing cryopreserved sperm. *Genetics,* 68:599–601.

Leigh, Egbert G., Jr. 1970. Natural selection and mutability. *Amer. Naturalist,* 104:301–305.

Levin, Bruce R. 1971. The operation of selection in situations of interspecific competition. *Evolution,* 25:249–264.

Levin, Donald A. and Harold W. Kerster. 1973. Assortative pollination for stature in *Lythrum salicaria. Evolution,* 27:144–152.

———, ———, Marianne Niedzlek. 1971. Pollinator flight directionality and its effect on pollen flow. *Evolution,* 25:113–118.

Levins, Richard. 1967. Theory of fitness in a heterogeneous environment. vi. The adaptive significance of mutation. *Genetics,* 56:163–178.

———. 1968. *Evolution in changing environments.* Princeton, Princeton Univ. Press, *ix* + 120 pp.

———. 1970. Extinction. In: *Mathematics and biology* (pp. 75–107). Providence, Amer. Mathematical Soc., *vii* + 156 pp.

Lewis, D. 1942. The evolution of sex in flowering plants. *Biol. Rev.* (Cambridge), 17:46–67.

——— and Leslie K. Crowe. 1956. The genetics and evolution of gynodioecy. *Evolution,* 10:115–125.

Ligny, Wilhelmina de. 1969. Serological and biochemical studies on fish populations. *Oceanogr. Mar. Biol., Ann. Rev.,* 7:411–513.

Liley, N. R. 1966. Ethological isolating mechanisms in four sympatric species of poeciliid fishes. *Behaviour,* 13 (Suppl.), *vi* + 197 pp.

Lloyd, Monte and Henry S. Dybas. 1966a. The periodical cicada problem. i. Population ecology. *Evolution,* 20:133–149.

——— and ———. 1966b. The periodical cicada problem. ii. Evolution. *Ibid.,* 20:466–505.

BIBLIOGRAPHY

McGillivray, M. E. 1972. The sexuality of *Mysus persicae* (Sulzer), the green peach aphid, in New Brunswick (Homoptera:Aphida). *Canadian J. Zool.*, 50:469–471.

McNeilly, T. and A. D. Bradshaw. 1968. Evolutionary processes in populations of copper tolerant *Agrostis tenuis*. *Evolution*, 22:108–118.

Mahendra, Beni Charan and Surendra Sharma. 1955. Classification of the modes of animal reproduction. *Ann. Zool.* (Agra), 1:85–86.

Manion, Patrick J. 1972. Fecundity of the sea lamprey (*Petromyzon marinus*) in Lake Superior. *Trans. Amer. Fish. Soc.*, 101:718–720.

Marty, Ju. Ju. 1965. Drift migrations and their significance to the biology of food fishes of the North Atlantic. *Intern. Council Northw. Atlantic Fish., Spec. Publ.*, 6:355–361.

Maslin, T. Paul. 1968. Taxonomic problems in parthenogenetic vertebrates. *Systematic Zool.*, 17:219–231.

Mather, Kenneth. 1940. Outbreeding and the separation of the sexes. *Nature*, 145:484–486.

Mathur, Daya S. 1962. Seasonal variations in the ovary and testes of *Barbus stigma* (*Puntius sophore*). *Zoologica Poloniae*, 12:131–144.

May, A. W. 1967. Fecundity of Atlantic cod. *J. Fish. Res. Board Canada*, 24:1531–1551.

Maynard Smith, John. 1964. Group selection and kin selection. Nature, 201:1145–1147.

———. 1966. *The theory of evolution*. Baltimore, Penguin, 336 pp.

———. 1968A. Evolution in sexual and asexual populations. *Amer. Naturalist*, 102:469–473.

———. 1968B. Haldane's dilema and the rate of evolution. Nature, 219:1114–1116.

———. 1970. Genetic polymorphism in a varied environment. *Amer. Naturalist*, 104:487–490.

———. 1971A. What use is sex? *J. Theoretical Biol.*, 30:319–335.

184

BIBLIOGRAPHY

————. 1971B. The origin and maintenance of sex. In: *Group selection* (pp. 163–175), G. C. Williams, editor. Chicago, Aldine-Atherton, 210 pp.

Mayr, Ernst. 1962. Accident or design, the paradox of evolution. In: *The evolution of living organisms* (pp. 1–14). G. W. Leeper, editor, Melbourne, Melbourne Univ. Press, *ix* + 459 pp.

————. 1963. Animal species and evolution. Cambridge, Mass., Harvard Univ. Press, *xiv* + 797 pp.

Mockford, Edward L. 1971. Parthenogenesis in psocids (Insecta: Psocoptera). *Amer. Zool.,* 11:327–339.

Møller, Dag. 1969. The relationship between Arctic and coastal cod in their immature stages illustrated by frequencies of genetic characters. *FiskDir. Skr. Ser. Havunders.,* 15:220–233.

Moorhead, Paul S. and Martin M. Kaplan (editors). 1967. Mathematical challenges to the Neo-Darwinian interpretation of evolution. *Wistar Inst. Symp. Monogr.,* 5, *ix* + 140 pp.

Moser, H. Geoffrey. 1967. Reproduction and development of *Sebastodes paucispinus* and comparison with other rockfishes off southern California. *Copeia* (4): 773–797.

Muller, H. J. 1932. Some genetic aspects of sex. *Amer. Naturalist,* 66:118–138.

Murdoch, W. W. 1966. Population stability and life history phenomena. *Amer. Naturalist,* 100:5–11.

Myrberg, Arthur A., Jr. 1965. A descriptive analysis of the behavior of the African cichlid fish, *Pelmatochromis guentheri* (Sauvage). *Animal Behavior,* 13:312–329.

Nei, Masatoshi. 1971. Fertility excess necessary for gene substitution in regulated populations. *Genetics,* 68:169–184.

Noble, Elmer R. and Glen A. Noble. 1961. Parasitology. *The biology of animal parasites.* Philadelphia, Lea and Febiger, 767 pp.

Noble, G. Kingsley. 1931. *The biology of amphibia.* New York, Dover Reprint (1954), 577 pp.

185

BIBLIOGRAPHY

Noble, Richard L. 1972. Mortality rates of walleye fry in a bay of Oneida Lake, New York. *Trans. Amer. Fish. Soc.,* 101:720–722.

Norden, Carroll R. 1967. Age growth and fecundity of the ale-wife, *Alosa pseudoharengus* (Wilson), in Lake Michigan. *Trans. Amer. Fish. Soc.,* 96:387–393.

Nur, Uzi. 1970. Evolutionary rates of models and mimics in Batesian mimicry. *Amer. Naturalist,* 104:477–486.

Nygren, A. 1966. Apomixis in the angiosperms, with special reference to *Calamagrostis* and *Poa.* In: *Reproductive biology and taxonomy of vascular plants* (pp. 131–140). J. G. Hawkes, editor, New York, Pergamon, 183 pp.

O'Donald, Peter. 1971. Natural selection for quantitative characters. *Heredity,* 27:137–153.

————. 1972. Natural selection of reproductive rates and breeding times and its effect on sexual selection. *Amer. Naturalist,* 106:368–379.

Oliver, James H., Jr. 1971. Parthenogenesis in mites and ticks (Arachnida: Acari). *Amer. Zoologist,* 11:283–299.

Olsen, M. W. 1965. Twelve year summary of selection for parthenogenesis in Beltsville Small White turkeys. *British Poultry Sci.,* 6:1–6.

Orians, Gordon H. 1969. On the evolution of mating systems in birds and mammals. *Amer. Naturalist,* 103:589–603.

Ornduff, Robert. 1971. The reproductive system of *Jepsonia heterandra. Evolution,* 25:300–311.

Otsu, Tamio and Richard J. Hansen. 1962. Sexual maturity and spawning of the albacore in the central South Pacific Ocean. *Fish. Bull., U.S.,* 62:151–161.

Pantelouris, E. M., A. Arnason, F.-W. Tesch. 1971. Genetic variation in the eel. III. Comparisons of Rhode Island and Icelandic populations. Implications for the Atlantic eel problem. *Marine Biol.,* 9:242–249.

Parker, G. A., R. R. Baker, V. G. F. Smith. 1972. The origin and evolution of gamete dimorphism and the male-female phenomenon. *J. Theoretical Biol.,* 36:529–553.

Prescott, G. W. 1968. *The algae: A review.* Boston, Houghton Mifflin, *xi* + 436 pp.

Raper, John R. 1966. *Genetics of sexuality in higher fungi.* New York, Ronald, *viii* + 283 pp.

Remane, Adolf and Carl Schlieper. 1971. *Biology of brackish water.* New York, Wiley Intersicence, *viii* + 372 pp.

Ricker, W. E. 1954. Stock and recruitment. *J. Fish. Res. Board Canada,* 11:559–623.

Ricklefs, Robert E. 1968. On the limitation of brood size in passerine birds by the ability of adults to nourish their young. *Proc. Nat. Acad. Sci.,* 61:847–851.

Risser, Paul G. 1970. Competitive responses of *Bouteloua curtipendula* (Michx.) Torr. *Amer. Midland Naturalist,* 84:259–262.

Rollins, Reed C. 1967. The evolutionary fate of inbreeders and nonsexuals. *Amer. Naturalist,* 101:343–351.

Rose, S. Meryl. 1959. Failure of survival of slowly growing members of a population. *Science,* 129:1026.

Ross, M. A. and John L. Harper. 1972. Occupation of biological space during seedling establishment. *J. Ecology,* 60:77–88.

Ross, M. D. and R. F. Shaw. 1971. Maintenance of male sterility in plant populations. *Heredity,* 26:1–8.

Sager, G. R. and John L. Harper. 1960. Factors affecting the germination and early establishment of plantains (*Plantago lanceolata, P. media,* and *P. major*). *British Ecol. Soc. Symp.,* 1:236–245.

Salisbury, E. J. 1936. Natural selection and competition. *Proc. Roy. Soc. London,* 121:47–49.

―――. 1942. *The reproductive capacity of plants.* London, Bell and Sons, *xi* + 244 pp.

Sastry, Akella N. and Norman J. Blake. 1971. Regulation of gonad growth in the bay scallop, *Aequipecten irradians* Lamarck. *Biological Bull.,* 140:274–283.

Savage, Jay M. 1968. The dendrobatid frogs of Central America. *Copeia* (4):745–776.

Schaffer, H. E. 1970. Survival of mutant genes as a branching process. In: *Mathematical topics in population genetics* (pp. 317–336), K. Kojima, editor, New York, Springer-Verlag, *x* + 400 pp.

Scheltema, R. S. 1971. The dispersal of the larvae of shoal-water benthic invertebrate species over long distances by ocean currents. *European Mar. Biol. Symp.,* 4:7–28.

Schmidt, Johannes. 1917. *Zoarces viviparus* L. and local races of the same. *Comptes rendus Lab. Carlsberg (Ser. Physiol.),* 13:279–397.

————. 1919. Experiments with *Lebistes reticulatus* (Peters) Regan. *Ibid.,* 14(5):1–8.

————. 1925. The breeding places of the eel. *Smithsonian Rept. (1924)*:279–316 + 7 pls.

Schopf, Thomas J. M. and James L. Gooch. 1971. Gene frequencies in a marine ectoproct: A cline in natural populations related to sea temperature. *Evolution,* 25:286–289.

Schultz, R. Jack. 1971. Special adaptive problems associated with unisexual fishes. *Amer. Zoologist,* 11:351–360.

Selander, Robert K. 1965. On mating systems and sexual selection. *Amer. Naturalist,* 99:129–141.

————, W. Grainger Hunt, Suh Y. Yang. 1969. Protein polymorphism and genic heterozygosity in two European subspecies of the house mouse. *Evolution,* 23:379–390.

————, Suh Y. Yang, Richard C. Lewontin, Walter B. Johnson. 1970. Genetic variation in the horseshoe crab (*Limulus polyphemus*), a phylogenetic "relic." *Ibid.,* 24:402–414.

Sette, Oscar E. 1961. Problems in fish population fluctuations. *Calif. Coop. Oceanic Fish. Invest., Rept.,* 8:21–31.

Simberloff, Daniel S. and Edward O. Wilson. 1969. Experimental zoogeography of islands: The colonization of empty islands. *Ecologyn* 50:278–296.

Simpson, A. C. 1951. Fecundity of plaice. *Fish. Invest. (London),* (2)17(5):1–27.

BIBLIOGRAPHY

Simpson, G. G. 1950. The meaning of evolution. New Haven, Yale Univ. Press. *xv* + 364 pp.

————. 1953. The major features of evolution. Columbia Univ. Press, New York and London, *xx* + 434 pp.

Skellam, J. G. 1955. The mathematical approach to population dynamics. In: *The numbers of men and animals* (pp. 31–46). J. B. Cragg and N. W. Pirie, editors, London, Oliver and Boyd, *viii* + 152 pp.

Sneath, Peter H. A. and Robert R. Sokal. 1973. Numerical taxonomy. W. H. Freeman and Company, San Francisco, *xv* + 573 pp.

Solbrig, Otto T. 1971. The population biology of dandelions. *Amer. Scientist,* 59:686–694.

————. 1972. Breeding system and genetic variation in *Leavenworthia. Evolution,* 26:155–160.

Sorensen, Frank. 1969. Embryonic genetic load in coastal Douglas-fir, *Pseudotusga Menziessi* var. *Menziesii. Amer. Naturalist,* 103:389–398.

Southwood, T. R. E. 1967. The interpretation of population change. *J. Animal Ecology,* 36:519–529.

Stalker, Harrison D. 1956. On the evolution of parthenogenesis in the Lonchoptera (Diptera). *Evolution,* 10:345–359.

Stebbins, G. Ledyard. 1950. *Variation and evolution in plants.* New York, Columbia Univ. Press, *xix* + 643 pp.

————. 1957. Self fertilization and population variability in the higher plants. *Amer. Naturalist,* 91:337–354.

————. 1960. The comparative evolution of genetic systems. In: *Evolution after Darwin,* vol. 1 (pp. 197–226), Sol Tax, editor, Chicago, Univ. Chicago Press, *viii* + 629 pp.

————. 1970. Variation and evolution in plants: Progress during the past twenty years. In: *Essays in evolution and genetics in honor of Theodosius Dobzhansky* (pp. 173–208), M. K. Hecht and W. C. Steere, editors, New York, Appleton-Century-Crofts, *xv* + 594 pp.

Stebbins, Robert C. and John R. Hendrickson. 1959. Field

studies of amphibians in Colombia, South America. *Univ. Calif. Publ. Zool.*, 56:497–540.

Stephenson, John. 1930. The Oligochaeta. Oxford, Oxford Univ. Press, *xvi* + 978 pp.

Stern, W. R. 1965. The effect of density on the performance of individual plants in subterranean clover swards. *Australian J. Agric. Res.*, 16:541–555.

Stewart, Frank M. and Bruce R. Levin. 1973. Partitioning of resources and the outcome of interspecific competition: A model and some general considerations. *Amer. Naturalist*, 107:171–198.

Strawn, Kirk. 1958. Life history of the pigmy seahorse, *Hippocampus zosterae* Jordan and Gilbert, at Cedar Key, Florida. *Copeia* (1):16–22.

Suomalainen, Esko. 1962. Significance of parthenogenesis in the evolution of insects. *Ann. Rev. Entomol.*, 7:349–366.

————. 1969. Evolution in parthenogenetic Curculionidae. *Evolutionary Biol.*, 3:261–296.

Sved, J. A. 1968. Possible rates of gene substitution in evolution. *Amer. Naturalist*, 102:283–293.

Takai, T. and A. Mizokami. 1959. On the reproduction, eggs, and larvae of the pipefish, *Syngnathus schlegeli* Kaup. *J. Shimonoseki College Fisheries*, 8:85–89.

Tamarin, Robert H. and Charles J. Krebbs. 1969. *Microtus* population biology. ii. Genetic changes at the transferrin locus in fluctuating populations of two vole species. *Evolution*, 23:183–211.

Tax, Sol and Charles Callender, editors. 1960. *Evolution after Darwin*, vol. 3, *Issues in evolution*. Chicago, Univ. Chicago Press, *viii* + 310 pp.

Thorson, Gunnar. 1950. Reproductive and larval ecology of marine bottom invertebrates. *Biol. Rev.* (Cambridge), 25:1–45.

Tomlinson, J. T. 1968. Improper use of the word bisexual. *Systematic Zool.*, 17:212.

BIBLIOGRAPHY

Trivers, Robert L. 1972. Parental investment and sexual selection. In: *Sexual selection and the descent of man, 1871–1971* (pp. 136–179), B. Campbell, editor, Chicago, Aldine-Atherton, *x* + 378 pp.

Turesson, Göte. 1922. The genetical response of the plant species to the habitat. *Hereditas*, 3:211–350.

Turner, John R. G. 1967A. Mean fitness and the equilibria in multilocus polymorphisms. *Proc. Roy. Soc. London* (B) 169:31–58.

———. 1967B. On supergenes. I. The evolution of supergenes. *Amer. Naturalist*, 101:195–221.

———. 1967C. Why does the genotype not congeal? *Evolution*, 21:645–656.

———. 1970. Some properties of two locus systems with epistasis. *Genetics*, 64:147–155.

———. 1971. Wright's adaptive surface, and some general rules for equilibria in complex polymorphisms. *Amer. Naturalist*, 105:267–278.

Tyler, Albert. 1967. Problems and procedures of comparative gametology and syngamy. In: *Fertilization*, vol. 1 (pp. 1–26). C. B. Metz and A. Monroy, editors, New York, Academic, *xiii* + 489 pp.

Uzzell, Thomas. 1970. Meiotic mechanisms of naturally occurring unisexual vertebrates. *Amer. Naturalist*, 104:433–445.

Valdeyron, G., B. Dommée, A. Valdeyron. 1973. Gynodioecy: Another computer simulation model. *Amer. Naturalist*, 107:454–459.

Van der Pijl, L. 1969. *Principles of dispersal in higher plants*. New York, Springer-Verlag, *viii* + 156 pp.

Van Valen, Leigh. 1965. Is there a genetic elite? *Amer. Naturalist*, 99:125–126.

Verner, Jared. 1964. Evolution of polygamy in the long-billed marsh wren. *Evolution*, 18:252–261.

Vladykov, Vadim D. 1964. Quest for the true breeding area

of the American eel (*Anguilla rostrata* LeSueur). *J. Fish Res. Board Canada,* 21:1523–1530.

Voipio, Paavo. 1969. Some ecological aspects of polymorphism in the red squirrel *Sciurus vulgaris* L. in northern Europe. *Oikos,* 20:101–109.

Waddington, C. H. 1957. *The strategy of the genes.* London, Allen and Unwin, Ltd., *ix* + 262 pp.

———. 1959. Evolutionary adaptation. *Perspectives Biol. Medicine,* 2:379–401.

Walberg, Charles H. 1972. Some factors associated with fluctuations in year-class strength of sauger, Lewis and Clarke Lake, South Dakota. *Trans. Amer. Fish. Soc.,* 101:311–316.

Wallace, Bruce. 1970. *Genetic load.* Englewood Cliffs, N.J., Prentice-Hall, *xi* + 116 pp.

Warburton, Frederick E. 1967. Increase in the variance of fitness due to selection. *Evolution,* 21:197–198.

Watson, D. M. S. 1936. A discussion of the present state of the theory of natural selection. *Proc. Zool. Soc. London,* 121(B):43–45.

Weismann, August. 1889. The significance of sexual reproduction in the theory of natural selection. In: *Essays upon heredity and kindred biological subjects* (pp. 254–338). Oxford, Oxford Univ. Press, *x* + 455 pp.

Wenner, Charles A. 1972. Aspects of the biology and systematics of the American eel, *Anguilla rostrata* (LeSueur). Dissertation, College of William and Mary, Williamsburg, Viriginia.

Wet, J. M. J. de. 1968. Diploid-tetraploid-haploid cycles and the origin of variability in *Dichanthium* agamospecies. Evolution, 22:394–397.

White, J. and John L. Harper. Correlated changes in plant size and number in plant populations. *J. Ecology,* 58:467–485.

White, M. J. D. 1970. Heterozygosity and genetic polymorphism in parthenogenetic animals. In: *Essays in evolution and genetics in honor of Theodosius Dobzhansky* (pp. 237–262), M. K. Hecht and W. C. Steere, editors, New York, Appleton-Century-Crofts, *xv* + 594 pp.

Williams, Austin B. 1969. A ten-year study of meroplankton in North Carolina estuaries: Cycles of occurrence among penaeidean shrimps. *Chesapeake Science,* 10:36–47.

Williams, George C. 1966A. *Adaptation and natural selection.* Princeton, Princeton Univ. Press, *x* + 307 pp.

————. 1966B. Natural selection, the costs of reproduction, and a refinement of Lack's principle. *Amer. Naturalist,* 100:687–690.

————, Richard K. Koehn, Jeffry B. Mitton. 1973. Genetic differentiation without isolation in the American eel, *Anguilla rostrata. Evolution,* 27:192–204.

———— and Jeffry B. Mitton. 1973. Why reproduce sexually? *J. Theoretical Biol.,* 39:545–554.

Wilson, Edward O. 1968. The ergonomics of caste in the social insects. *Amer. Naturalist,* 102:41–66.

Wright, John W. and Charles H. Lowe. 1968. Weeds, polyploids, parthenogenesis, and the geographical and ecological distribution of all-female species of *Cnemidophorus. Copeia* (1):128–138.

Wright, Sewall. 1956. Modes of selection. *Amer. Naturalist,* 90:5–24.

Wydoski, Richard S. and Edwin L. Cooper. 1966. Maturation and fecundity of brook trout from infertile streams. *J. Fish. Res. Board Canada,* 23:623–649.

Wynne-Edwards, V. C. 1962. Animal dispersion in relation to social behavior. London, Oliver and Boyd, *xi* + 653 pp.

Zweifel, Richard G. 1965. Variation and distribution of the unisexual lizard *Cnemidophorus tsselatus. Amer. Museum Novitates,* 2235:1–49.

193

Index

adaptive performance, 48, 75, 76, 77ff. *See also* fertility, fitness
adultery, 128, 130
age and sexual attractiveness, 128
age groups, genetic differences, 86, 87
Allard, R. W., 160, 161
amphibians, 105, 133
anisogamy vs. isogamy, 113–116
Aphid-Rotifer Model, 12, 15ff, 116; compared to other models, 22, 33, 35, 36, 37, 44; for protists, 112; for vertebrates, 105–106
aphids, 11, 15, 16, 106, 118
armadillo, 103, 115
Armitage, K. B., 132
asexual reproduction, advantages of, 111, 112, 145, 160–162, 167–169; classification, 114–117; defined, 4, 114, 115; evolutionary loss of, 41ff, 50ff, 102; requirements for, 103, 111. *See also* sexual reproduction
assortative mating, 53ff
Asterias, 44
Australia, 52

Bacci, G., 22, 118
Baker, R. R., 111ff, 120
Bateson, G., 169
bdelloid rotifers, 163
Beardmore, J. A., 25
beauty, in relation to age, 128
biotic evolution, 155
bipinnaria, 49
birds, as low-fecundity organisms, 102; eggs of, 103; reproductive costs, 126; with reversed sexual roles, 134
bisexual, 115
blennies (fishes), 83
Bodmer, W. F., 91, 145, 146
Bonner, J. T., 3, 4, 7
Boorman, S. A., 156

Bossert, W. H., 117, 125
brackish water, 164
Bradshaw, A. D., 88, 89
budding vs. parthenogenesis, 10

canalization, of character optima, 91, 93, 94; of fitness, 68, 70, 71; and genetic dominance, 109; of reproductive functions, 130; of visual mechanisms, 47
care of young, male vs. female strategies, 134
Carson, H., 105
Chlora, 95
cicada, 117
cichlid fishes, 136ff
cladogenesis, random, 165–166
Clausen, J., 92
clones, competition between, 17, 18, 29ff, 34, 62; size of territory, 28, 34
Cnemidophorus, 161, 162
cod, adult movements, 54; blood types, 82, 83; fertility vs. age, 98; fertility excess, 67, 68; intensity of selection, 65; larval ecology, 44, 45, 54, 66, 83; local differentiation, 82, 83, 84; selectivity of death, 63, 64; stock-recruitment relation, 72, 73
Cod-Starfish Model, 52ff, 83; compared to other models, 59
coelenterates, 5, 11, 52
Cohen, J., 6
cohort half-life, 64, 67
colonizing species, frequency of selfers and asexuals, 160ff
comparative evidence, 3, 7–8, 134
competition between clones, 17, 18, 29ff, 34, 62
competition in juvenile stages, 36ff
competitive exclusion, 18, 29, 30

195

Library of Congress Cataloging in Publication Data

Williams, George Christopher, 1926–
    Sex and evolution.

    (Monographs in population biology, 8)
    Bibliography: p
    1. Evolution. 2. Sex (Biology) I. Title.
II. Series. [DNLM: 1. Evolution. 2. Reproduction.
3. Sex. W1 M0568L v. 8/QH471 W723s]
QH371.W54   575   74-2985
ISBN 0-691-08147-6
ISBN 0-691-08152-2 (pbk.)